SpringerBriefs in Molecular Science

Chemistry of Foods

Series editor

Salvatore Parisi, Industrial Consultant, Palermo, Italy

More information about this series at http://www.springer.com/series/11853

Caterina Barone · Luciana Bolzoni
Giorgia Caruso · Angela Montanari
Salvatore Parisi · Izabela Steinka

Food Packaging Hygiene

 Springer

Caterina Barone
ENFAP Comitato Regionale Sicilia
Palermo
Italy

Luciana Bolzoni
Experimental Station for the Food
  Preserving Industry (SSICA)
Parma
Italy

Giorgia Caruso
Industrial Consultant
Palermo
Italy

Angela Montanari
Experimental Station for the Food
  Preserving Industry (SSICA)
Parma
Italy

Salvatore Parisi
Industrial Consultant
Palermo
Italy

Izabela Steinka
Medical University of Gdansk
Gdansk
Poland

ISSN 2191-5407   ISSN 2191-5415   (electronic)
SpringerBriefs in Molecular Science
ISSN 2199-689X   ISSN 2199-7209
Chemistry of Foods
ISBN 978-3-319-14826-7   ISBN 978-3-319-14827-4   (eBook)
DOI 10.1007/978-3-319-14827-4

Library of Congress Control Number: 2014959822

Springer Cham Heidelberg New York Dordrecht London

Printed on acid-free paper

Springer International Publishing AG Switzerland is part of Springer Science+Business Media (www.springer.com)

Contents

Chapter 1
The Influence of the Chemical Composition of Food Packaging Materials on the Technological Suitability: A Matter of Food Safety and Hygiene

Salvatore Parisi, Caterina Barone and Giorgia Caruso

Abstract Normally, the so-called Declaration of Food Contact Compliance is one of the most known and debated argumentations with reference to food packaging materials. This topic has been extensively discussed in the last years. However, another aspect remains to be shown and critically analysed: the 'technological suitability' for food applications. By the viewpoint of the European Legislator, this concept is the second requirement for the safe and legal use of food containers. On the other hand, the definition of technological suitability is not available in existing official norms or in most known food quality standards, while a specific statement has been recently made in the scientific literature. Secondly, technological suitability should be necessarily linked and influenced by different and known factors: the chemical composition of the food container or food contact material; the technological classification of the container; the chemical profile of the packaged food; the production and packaging process; and the problem of correct storage procedures for food packaging materials and packaged products. This work would show several practical applications with reference to the connection between the chemical composition of food packaging materials and the predictable behaviour of the container in 'normal conditions'.

Keywords Chemical risk · Declaration of compliance · Food hygiene · Food packaging · Packaging failures · Technological suitability

Abbreviations

BRC	British Retail Consortium
BADGE	Bisphenol A diglycidyl ether
BFDGE	Bisphenol F diglycidyl ether
DEHP	Bis(2-ethylhexyl) phthalate
BSI	British Standards Institution
CAS	Chemical Abstract Service

C. Barone et al., *Food Packaging Hygiene*, Chemistry of Foods,
DOI 10.1007/978-3-319-14827-4_1

DoC	Declaration of Food Contact Compliance
DPB	Dibutyl phthalate
DIBP	Diisobutyl phthalate
DIPN	Diisopropyl naphthalene
FRF	Fat consumption reduction factor
FWA	Fluorescent whitening agent
FQMS	Food quality management system
FPP	Food packaging producer
FPM	Food packaging material
FP	Food producer
FQMS	Food quality management system
EU	European Union
GSFS	Global Standard for Food Safety
GMP	Good manufacturing practices
HACCP	Hazard analysis and critical control points
IoP	Institute of Packaging
IFP	Integrated food product
IFS	International Featured Standards
ITX	2-Isopropyl thioxantone
MOSH	Mineral oil saturated hydrocarbon
MOAH	Mineral oil aromatic hydrocarbon
NOGE	Novolac glycidyl ether
OML	Overall migration limit
PCP	2,3,4,5,6-Pentachlorophenol
PCB	Polychlorobyphenyl
PAH	Polycyclic aromatic hydrocarbon
PAA	Primary aromatic amine
PAS	Publicly Available Standard
SVOC	Semivolatile organic compound
SML	Specific migration limit
USA	United States of America
VOC	Volatile organic compound

1.1 Food Safety and Packaging Materials

The connection between food safety and food packaging materials (FPM) is one of the most debated arguments in the modern world of food production. From a general viewpoint, the commercial impact of every new law or regulation concerning FPM can be easily expected. On the other hand, it should be demonstrated that food producers (FPs) are completely able to manage FPM and its peculiar features from the technical viewpoint: the world of FPM may appear often 'indecipherable' for food manufacturers [1, 2].

Consequently, the role of food packaging is one of the most discussed topics today. This 'accessory' but fundamental material is constantly considered in the whole chain of food and feed commodities: farming, production, distribution, retail and catering. However, every different player in the food chain seems to consider FPM in a different way.

FPM is generally seen as a sort of accessory structure by inexperienced subjects with reference to the real edible content [3]. On these bases, it could be inferred that FPs are not responsible for the use of FPM and related consequences: in fact, containers are clearly non-edible! However, the current legislation on food hygiene and safety has slowly but increasingly modified the original viewpoint about FPM in the last 30 years. Nowadays, the matter of food packaging is considered one of essential bases of the modern food safety strategy worldwide. In detail, this process has been observed in most part of the European countries, although several different and independent strategies have been elaborated. For example, the Italian legislation had initially proposed a complex and original series of norms about FPM in the 1970s [4]. Anyway, the harmonisation of different national legislations in the European Union (EU) has finally produced a coordinated system of common standards for the production and the commerce of food commodities [2]. This process has been observed for FPM also, although several peculiarities can be mentioned at present in relation to EU food legislation.

From a general viewpoint, the Regulation (EC) No. 1935/2004 has finally placed FPM on the same level of the edible content in spite of its clearly different origin and nature. Actually, this document corresponds to one of practical applications of the previous Regulation (EC) No. 178/2002 of the European Parliament and of the Council of 28 January 2002, with reference to food safety.

In detail, the Regulation (EC) No. 1935/2004 has clearly stated that packaging materials are active components of the so-called integrated food product (IFP) when used by FP [5]. In other words, the FP is surely responsible for its own product, including the use and the management of food containers and similar components. These packaging materials are certainly able to determine and influence the safety and integrity of the packaged product with distinctive advantages, but the possibility of damages for the consumer has to be considered at the same time.

For this and other reasons, the creation and the implementation of adequate 'good manufacturing practices' (GMP) by food packaging producers (FPP) is surely welcomed and requested by current food quality standards. With relation to food products, similar requirements are recommended by several of the most known and considered protocols: two examples are surely the Global Standard for Food Safety (GSFS) by the British Retail Consortium (BRC) and the International Featured Standards (IFS) Food. However, these protocols are specifically addressed to the world of the food production.

On the other hand, the role of FPP appears dissimilar from the position of FP. In relation to FPP, the Regulation (EC) No. 1935/2004 has introduced the concept of GMP. Subsequently, the Publicly Available Standard (PAS) 223:2011 has been prepared by the British Standards Institution (BSI) with the aim of specifying basic requirements for prerequisite programmes to assist in controlling food safety hazards [6]. This document has been specifically created for every FPP plant with the necessity of meeting the requirements specified by the BS EN ISO 22000 norm.

In other words, PAS 223:2011 allows the creation and the implementation of a food quality management system (FQMS) near FPP. This approach—the creation and implementation of adequate GMP—corresponds to the practical and synergistic application of different codes of practices. For example, the BS EN 15593 norm—management of hygiene in the production of packaging for foodstuffs, by BSI—can be cited here. Another useful document is the 'Recommended international code of practice—General Principles of Food Hygiene' [7].

Apparently, this matter seems circumscribed and clarified enough: the abundance of scientific literature and regulatory norms about the safety of FPM should corroborate this reflection. However, many points remain 'obscure'. In fact, every IFP—the synergic sum of the following terms: food, FPM and other accessory and 'invisible' services as the so-called quality control—shows different features depending on various factors. Two of these parameters are surely the nature and the commercial typology of FPM, according to several authors [2, 8]. However, the role of FPM on IFP performances appears often unclear because of the concomitant presence of other factors. In relation to food alterations, the influence of incorrect food storage conditions is well known, but defective FPM may reduce IFP performances at the same time in a similar way.

On the other hand, FP and PPP may consider the IFP performance by different viewpoints. Naturally, the safety and integrity of IFP are compulsory objectives for FP [1, 9]. The same thing has to be affirmed by FPP. However, the concept of 'damage' for FPP is generally coincident with the difference between the expected performance of FPM when used to preserve foods and the real behaviour of the final IFP. For example, a particular graphic failure on the surface of packaged foods can be extremely meaningful for FPP, while the same defect may appear without consequences for FP. Naturally, the opposite situation can also occur: IFP may appear sensorially damaged (differences of colour, shape, apparent texture) without real food safety risks because of light packaging imperfections [1]. It has to be recognised that this alarming reaction is generally monitored by mass retailers on the basis of customer complaints. In other words, the opinion of the final customer is important and may be coincident with the judgement of the 'final user' of FPM.

The viewpoint of the last player of the food chain (with the exclusion of final customers) is significant [1]. Every defect is always important for the final distributor: this subject considers all possible IFP failures on the same level, without distinction between primary causes (food, packaging, label, storage conditions, transportation, etc.). For this reason, most part of international mass retailers have increasingly chosen a restricted number of FP for the management of their 'private label' IFP. These food companies have to comply with well-known food quality standards (GSFS, IFS Food, the ISO 22000:2005 norm, etc.). The same thing is requested for FPP: at present, the most recognised quality standard for FPM is the BRC-IoP Global Standard for Packaging and Packaging Materials, by the BRC and the Packaging Society, formerly known as the Institute of Packaging (IoP).

On the other hand, the different nature of FPM in comparison with foods and beverages has to be noted. By the regulatory viewpoint, FPM is an important part of the IFP: additionally, mass retailers and quality auditors prefer to consider the position of FPM on the same level of the edible content. However, FPM seems to remain 'a world apart' for several professionals involved in the world of food production, management and surveillance: these materials are generally non-edible products with the exception of a few situations, while contained products are surely edible. Consequently, FPM appears technologically and inherently different from foods and beverages. For this and other reasons, the 'hazard analysis and critical control points' (HACCP) approach for the management of food risks may appear extraneous to the world of FPM.

In effect, the above-mentioned PAS 223:2011 and BRC-IoP documents have been created with the aim of introducing ISO 9001-based quality systems and the HACCP approach in the industry of FPM and related materials [2, 5]. Naturally, different competencies are needed and equally represented: food technology, food safety and hygiene (by the medical viewpoint), chemistry, microbiology, veterinary medicine, entomology, etc.

Finally, the role of health officers has to be considered. Because of different competencies, these professionals have to be adequately trained. In reference to FPM, the exiguity of sufficient information in the scientific literature has to be noted. However, health officers should be able to consider, evaluate and judge correctly FPM by the hygienic viewpoint: in effect, FPM is considered one of IFP parts [10]. On the other hand, is FP able to evaluate and manage FPM?

The 'right' strategy is correlated with five important but different viewpoints:

- The 'hygiene and safety' approach (this is the preferred ambit in the medical environment)
- The veterinary viewpoint
- The food technology approach
- The microbiological viewpoint (the correlated risk is the most known and debated in the HACCP ambit)
- And finally, the chemical approach.

Actually, the last viewpoint should be carefully considered because of the following reasons:

1. Every food or beverage has always its own chemical composition; on the other hand, two similar foods or beverages can surely have different chemical compositions. Consequently, there are different versions of the same IFP on the market
2. The preparation of foods and beverages can be influenced by various factors. One of these parameters is the composition of different raw materials: there are different versions of the same raw material on the market
3. Additionally, food additives and other chemicals can heavily influence the chemical composition and related sensorial features of the final IFP, including apparent properties of FPM
4. Finally, the presence of different molecules (small amounts) in food products with uncertain origin can be detected. Actually, these compounds may be derived by predictable chemical reactions: microbial fermentations, catalysed physical–chemical alterations and possible migration phenomena from FPM at the food/packaging interface.

As a result, the problem of chemical contamination should be carefully studied: adequate preventive or corrective actions should be taken in the ambit of food production, according to basic HACCP principles and the so-called quality vision [1]. This description is surely simplified, but the general idea of chemical contamination is often linked to migration episodes from FPM, instead of other food-related causes (examples: excessive quantity of undesired food additives and microbial fermentations with low probability). Actually, main pilasters of the above-mentioned HACCP approach are as follows [1]:

(a) The microbiological risk
(b) The chemical risk (detection of chemicals in small amounts with microscopic dimensions, including the presence of nanoparticles also)
(c) The physical risk (presence of foreign substances with macroscopic dimensions).

Microbiological and physical risks are not discussed here except for possible correlations with macroscopic or microscopic evidences of the chemical risk [10]. On the other side, the approach to the evaluation and the management of chemical hazards in the ambit of the food production is not simple at first sight.

Generally, the chemical risk is coincident with the concept of chemical contamination. In other words, the possible occurrence of apparent or clearly defined chemical hazards occurs if the designed IFP is not correlable with the planned chemical composition. One or more of the below-mentioned situations can occur:

1. Diffusion of foreign but edible contaminants in the inner and/or external layers, including the superficial area
2. Diffusion of foreign and non-edible contaminants in the inner and/or external layers, including the superficial area
3. Transformation of one or more original components of the final IFP because of predictable or unknown factors, with active influence of FPM
4. Transformation of one or more original components of the final IFP because of predictable or unknown factors under incorrect storage conditions, without active influence of FPM
5. Transformation of one or more original components of the final IFP because of predictable or unknown factors under incorrect storage conditions, with active influence of FPM
6. Apparent transformation of sensorial features because of predictable or unknown factors under normal or incorrect storage conditions, with or without FPM ruptures or other damages.

The above-mentioned list is not certainly exhaustive. Many additional phenomena may be included [1].

In reference to the connection between FPM and chemical hazards in food products, more research is surely needed. On the other hand, regulatory instruments seem to be adequate enough at present. This situation is observed in the EU and in the United States of America (USA) at least. Section 1.2 is dedicated to the European approach to the evaluation and the management of FPM-related chemical risks.

1.2 Regulatory Aspects in the EU: The Current Situation

The current EU regulatory system studies the world of FPM by the viewpoint of final users (FP, mass retailers, etc.) with the obvious exclusion of the final consumer [1].

With exclusive reference to FPM and food packaging objects—this definition comprehends every type of food contact substance, including permanent and temporary coatings for food-processing machinery and equipment—the Framework Regulation (EC) No. 1935/2004 may be still considered after 10 years as the best result in terms of the harmonisation between different national perspectives. Several common points have been finally established: one of these elements is related to the 'supporting documentation' for the DoC (Chap. 3). Essentially, this documentation is composed of chemical analyses, tests and other evaluations. Different from the interpretation of the past and repealed Directive 89/109/EEC, every written declaration stating that food packaging materials and objects comply with the rules applicable to them has to be supported by appropriate documentation [4].

As mentioned above, the (EC) Reg. No.1935/2004 introduces the obligatory DoC (Chap. 3): this document has to be supplied by FPP. In addition, the final user has to be able to evaluate the real suitability of FPM to the intended use on the basis of the DoC and the supporting documentation, before using them.

Actually, this requirement can be differently intended. For example, every final user, including food service providers, should test the real suitability of purchased plastic plates and other 'temporary' FPM, including biodegradable shoppers. In other words, the final user should evaluate the 'performance' of FPM with relation to the final IFP [10]. This discussion—the evaluation of the 'technological suitability' of FPM—is one of new frontiers because of the intrinsic meaning by the safety and hygiene viewpoint. In addition, the HACCP approach is fully connected with this matter because of the management of FPM as a basic component of IFP, according to the Regulation (EC) No. 178/2002 [4].

Anyway, it has to be repeated that FP and other users of FPM can use these containers and objects after the full evaluation of the DoC and the estimation of the 'technological suitability' [10]. With reference to this topic, several guidelines have recently expressed and clarified the basic concept of 'technological suitability' in Italy at least [4].

With relation to the specific problem of chemical hazards by FPM contamination, the most recent EU legislation is certainly the Regulation (EU) No. 10/2011 on plastic materials and articles intended to come into contact with food. In detail, this Regulation has rediscussed several topics of general interest with concern to the food contact compliance of plastic FPM (food packaging objects have also been cited). Probably, the most meaningful variation has concerned the suitability of simulative studies and alternative methods of analysis in comparison with official or recognised methods. In other words, alternative methods and simulations can be also used with the aim of demonstrating the compliance of FPM to food contact applications [4]. Once more, the importance of adequate 'supporting documentation' for the DoC appears basic.

As a result, authors have decided to discuss two of the above-mentioned topics in a simplified way. Section 1.1 concerns the nature of the DoC and related argumentations in relation to possible chemical contamination, while the argument of Sect. 1.4 is the technological suitability of FPM.

1.3 The Declaration of Food Contact Compliance

The DoC is constantly cited in the above-mentioned Regulations (EC) No. 1935/2004 and (EU) No. 10/2011 [4, 11]. This document has to be made available to the competent authorities on demand; in addition, it has to be attached to the FPM after its production as a paper or electronic document [12].

It has to be remembered that the final (also named downstream) user can be differently classified. There are three different classes of downstream users:

(a)   All food manufacturers
(b)   All food packers without processing activities
(c)   All distributive operators with possible packing activity, including catering services.

In fact, every FPP is obliged to write and supply the above-mentioned DoC for every produced FPM and/or single component. The responsibility of this player of the food chain is circumscribed to the redaction of this Declaration and the prompt availability of the supporting documentation. In reference to these data, the following documents may be considered [4, 12]:

• Recipes/process data/GMP documentation
• Processing data
• Test results
• Simulative studies
• Third-party certificates and analytical reports
• Risk assessment studies.

Additionally, the DoC does not release the downstream user from the exercise of 'due diligence' [11]. In fact, the final user is always responsible for the supplied IFP with concern to the safety and the legality. In detail, he is obliged to do the following:

1. Evaluate the compliance of FPM before using it, on the basis of the DoC and the supporting documentation and
2. Verify the suitability of FPM for the intended use before using it.

As a result, the position of the final user is surely important [1]. From the chemical viewpoint, what is the real meaning of the above-mentioned DoC and related supporting documentation? Nowadays, common consumers seem apparently persuaded that FPM are substantially unreactive with foods and beverages [4]. However, the concept of chemical interaction between packed food and

FPM is widely accepted in the scientific world [10] and the recent regulatory has evidenced the attention of the national legislator in a number of countries. Anyway, three conditions have to be respected in relation to the possible migration of chemical substances from FPM to packed foods (the inverse migration is always possible). The following can be affirmed [4]:

(a) The human safety cannot be compromised
(b) The chemical composition of packed foods cannot be modified in an unacceptable way in reference to the original product (these conditions state IFP and packed foods are two different concepts)
(c) Sensorial features of the IFP cannot be altered (for example, texture and colour are either correlable with packed foods and FPM at the same time).

As a result, the migration of potentially toxic or harmful substances from FPM to food products has to be carefully evaluated, with or without the modification of chemical compositions and sensorial features. Actually, every chemical or physical modification of IFP is important because of the intrinsic meaning of 'warning light': sometimes, food hygiene alerts may be highlighted by apparently strange or grotesque phenomena on the organoleptic viewpoint [1].

With specific relation to the diffusion of chemical substances from FPM to foods, two terms have to be considered: the 'overall' and the 'specific' migration. According to the Reg. (EU) No. 10/2011 (plastic FPM), the EU legislation has already defined the overall migration limit (OML). This quantity corresponds to the amount of substances that can be released by FPM to foods or food simulants: it cannot exceed 10 mg per square decimetre in the EU, with the exception for articles destined to contain foods for infants and small children. This limit means a sort of primary discriminating rule.

Subsequently, other specific migration limits (SML) have been defined for peculiar substances and in function of FPM. Once more, this is the situation for plastic FPM and similar articles. In addition, the introduction of the EU list of allowed additives for the manufacturing of plastic FPM and similar articles has to be signalled. This series of chemical substances, also defined the 'union list', contains several additives with a valid SML. More research is still needed. In fact, the IFP should be examined by different viewpoints: food processing, food packaging, food logistic and other factors should be investigated. For example, the influence of FPM on IFP should be estimated under normal storage conditions.

However, several foreign substances (mineral oils) have been recently found in packed foods because of the probable migration from the secondary packaging (carton board), in spite of the obvious presence of the primary FPM as 'barrier' [13]. This situation and other researches have alarmed the whole sector of food production and most part of paper and board FPP because more than 50 % of raw materials for this type of packaging originated from paper recycling. On these bases, EU and national authorities are still monitoring the problem, and the regulatory system is constantly evolving towards possible and sustainable measures [14]. Similar reasoning can be easily made in relation to food commodities that can be temporarily stored into warehouses, cargos, etc. [11].

With concern to the EU situation, most important and researched contaminants in paper and board FPM are listed below [4]:

- 2,3,4,5,6-pentachlorophenol (PCP), chemical formula: C_6HCl_5O, chemical abstract service (CAS) number: 87-86-5
- Phthalates: for example,
  - Dibutyl phthalate (DPB), chemical formula: $C_{16}H_{22}O_4$, CAS number: 84-74-2
  - Bis(2-ethylhexyl) phthalate (DEHP), chemical formula: $C_{24}H_{38}O_4$, CAS number: 117-81-7
  - Diisobutyl phthalate (DIBP), chemical formula: $C_{16}H_{20}O_4$, CAS number: 84-69-5
- Volatile organic compounds (VOC) and semi-volatile organic compounds (SVOC)
- Diisopropyl naphthalene (DIPN), a mixture of isomeric diisopropylnaphthalenes
- Polycyclic aromatic hydrocarbons (PAH)
- Formaldehyde, chemical formula: CH_2O, CAS number: 50-00-0
- Glioxal, chemical formula: $C_2H_2O_2$, CAS number: 107-22-2
- Polychlorobyphenyls (PCB)
- Primary aromatic amines (PAA)
- Fluorescent whitening agents (FWA)
- Antimicrobial substances
- Photoinitiators: for example,
  - Benzophenone, chemical formula: $C_{16}H_{20}O_4$, CAS number: 119-61-9
  - 4,4′-bis(dimethylamino) benzophenone (also named Michler's ketone), chemical formula: $C_{17}H_{20}N_2O$, CAS number: 90-94-8
- Bisphenol A, chemical formula: $C_{15}H_{16}O_2$, CAS number: 80-05-7
- Dioxins
- Mineral oil saturated hydrocarbons (MOSH) and mineral oil aromatic hydrocarbons (MOAH)
- Heavy metals: lead, cadmium and mercury
- Microbiological agents: yeasts and moulds.

Actually, this list is not exhaustive. Other chemicals may be added: the possibility of chemical reactions between original FPM chemical compounds and the food matrix should be recognised. Additionally, the presence of residuals of FPM manufacturing is always possible. The most known and recent situation concerns bisphenol A, while other endocrine-disrupting compounds have already 'obtained' their placement in the history of food contamination: bisphenol A diglycidyl ether (BADGE), bisphenol F diglycidyl ether (BFDGE) and novolac glycidyl ether (NOGE) [4, 15, 16].

With concern to recommended analytical methods and limits, the Reg. (EU) No. 10/2011 clarifies this point for plastic FPM [4]. Consequently, this Regulation can be certainly considered an useful regulatory instrument. In detail, Annex I mentions the list of authorised substances for the production of plastic articles: additionally, the list shows also the possible use for every chemical, the related SML (mg/kg), the possibility of correcting migration results by the fat consumption reduction factor (FRF) and other recommendations.

On the other side, the same EU Regulation has defined peculiar restrictions for several metallic elements in relation to the specific migration [4]. These metals are barium, iron, copper, lithium, cobalt, manganese and zinc. Moreover, restrictions have been applied to several PAA without mention in the above-mentioned Annex I (SML is defined 0.01 mg/kg for these substances).

In relation to analytical methods for the assessment of migration, the definition and the classification of test conditions is absolutely needed. The Annex III of the Regulation (EU) No. 10/2011 concerns specifically food simulants with the mention of their usage conditions and related criteria for the assessment of OML, while the Annex V concerns analytical procedures for the assessment of SML in accordance with requisites of the Regulation (EC) No. 882/2004. Actually, the comprehension of these annexed documents may be arduous for several players of the food chain and specifically for FP. This point has to be carefully discussed because the evaluation of DoC depends on the ability and chemical competencies of final users.

The full and detailed description of the Regulation (EU) No. 10/2011 and other similar EU norms is not the aim of this work; moreover, a notable part of FPM classes and typologies are differently managed, depending on the peculiar country and the correlated national legislation. Similar discussions should take more pages of this book; it can be predicted that other future books of this 'Chemistry of Foods' series will consider the topic in a more comprehensive way. In the meantime, the interested reader is cordially invited to search for more specific literature in selected references when speaking of OML and SML values in the EU [4]. For example, the EN 1186-1:2002 norm is a guideline for the selection of conditions and test methods for overall migration in relation to FPM, while EN 13130:2005 standard protocols and CEN/TS 13130:2006 methods concern SML values [4].

As stated above, several FPM sectors are differently managed in the EU, country by country. Additionally, the problem of recycled raw materials has to be considered. For example, the Italian legislation recommends strong surveillance on paper and board FPM with specific reference to PCB, lead, FWA and other analytes: dithiocamarbates, xanthogenates, trivalent chromium, primary and secondary aromatic amines, etc. Interested readers can surely find detailed guidelines in the EU ambit, but these documents are not legally compulsory: an useful example can be the Resolution AP (2002) 1 on paper and board materials and articles intended to come into contact with foodstuffs, by the Council of Europe [4]. Similar situations can be observed in the EU with concern to glass and metal FPM.

Moreover, there are different methods for the evaluation of sensorial properties of packed foods and beverages after packing. An interesting example is the Italian UNI 10192:2000 norm: this document concerns the evaluation of possible sensorial defects on foods after contact with FPM [4].

The so-called set-off (transfer of printing inks from FPM to foods) is another important example of sensorially estimable food contamination, when the failure is macroscopic [1].

In effect, the discussion of ghosting effects [10] can be helpful because of the possibility of introducing the matter of technological suitability of FPM. This matter may be seen as one of the connections between the complex of regulatory norms about FPM and the 'hygiene package' by means of the above-mentioned HACCP approach [1].

1.4 The Problem of the Technological Suitability of Food Packaging Materials

With exclusive relation to the EU legislation, FPM can be used on condition that [4, 5]

1. The above-mentioned DoC has been made available by FPP for every batch of produced FPM
2. The final user has really evaluated and examined the compliance of purchased FPM in relation to the real use
3. The final user has really evaluated the so-called technological suitability of purchased FPM.

In other words, these conditions have to be fully satisfied: in addition, it should be noted that the simple DoC is not sufficient for final users because they are not dispensed with the exercise of 'due diligence' [11].

Once more, main responsibilities are ascribed to the downstream user (FP), while the FPP is 'only' obliged to produce the related DoC. In fact, the food manufacturer or packer is always responsible for the supplied IFP with concern to the safety and the legality: He cannot share this responsibility with the FPP because the last player is supposed to create (design, produce, test) FPM on the basis of received information by the final user. Finally, the FP should verify that the obtained FPM is compliant with the intended use [1]. This evaluation is not circumscribed to the mere examination of printed DoC: most known food quality systems have already clarified this point. For example, the IFS Food standard, version 6, requires that the suitability of FPM has to be verified by final users for every relevant food product on the basis of HACCP studies (clause 4.5.4). Sensorial evaluations, storage tests, chemical analyses, and migration tests may be used for this evaluation [9].

In fact, every FPM may be intended as a sort of 'suit' for general applications: the real suitability has to be verified for every new food [1].

On these bases, the problem of the technological suitability may appear very 'thorny' because of the scarcity of related information and scientific literature [4, 10]. This requisite has not been defined in the EU with exclusive reference to FPM. On the other side, the same concept appears obvious and tacitly agreed on the ground of food safety: Regulations (EC) No. 178/2002 and the 'hygiene package' are good examples [4]. Anyway, there is not a clear definition of technological suitability on the regulatory ground, while above-mentioned food quality standards seem to recognise the problem without written specifications, except for the recommendation of possible testing methods. On the other hand, some national legislation has repeatedly considered the obligation. Anyway, technological suitability may be intended as the capability of FPM to show expected performances for 'intended' applications without deviations [10]. This concept is based on three important pilasters at least:

1. The technological suitability cannot be predicted or stated without the preventive communication of the predictable use by the final user to the FPP
2. The problem of the 'intended' use has to be considered. In other words, the performance of FPM has to be necessarily evaluated in contact with edible foods and under usual conditions
3. Finally, the technological suitability appears to be ideally extended until the end of the shelf life or the IFP.

Several reflections should be done before proceeding. First of all, the 'intended use' is absolutely essential: different situations can occur when the same FPM is used to obtain similar IFP with two or more different foods. Basically, at least three factors are needed before estimating the performance of the peculiar FPM on the final IPF:

• The food content
• The packing and processing system
• The predictable storage in terms of conditions (temperature, environmental locations, etc.).

In fact, all possible modifications of the IFP include the chemical composition and correlated sensorial properties of the packaged food or beverage. For example, the migration or organic contaminants from FPM to the food surface may be macroscopically evident in several situations, while other IFP may appear sensorially good or excellent. In reference to the last situation, the food safety may not be compromised, but hygiene concerns can be very evident to consumers.

Secondly, what is the real meaning of the 'predictable behaviour' of FPM in contact with foods? In fact, this concept should be differently intended if compared with the intended use, in spite of the apparent connection. A peculiar FPM may be used for packing similar or completely different foods: as a result, it may be inferred that the behaviour of the same FPM—actually, the performance of the resulting IFP—should theoretically be dissimilar for every application. The question is what is the degree of similarity between different IFP? Consequently, the intended use may determine dissimilar IFP performances or 'predictable behaviours', but this correlation is not sure and it should be continually validated. Section 1.5 shows several food applications with unpredictable results.

Additionally, the problem of the temporal deadline of the technological suitability should be briefly discussed. In effect, the shelf life of the IFP correspond to the real deadline for used FPM. However, every container and similar food packaging components have certainly their own expiration dates: these dates are established by FPP. Substantially, the technological suitability of FPM is temporarily dependent on FP until its use; on the other hand, this property is initially limited by FPP. As a result, the shelf life of the finished IFP cannot be correlated with the technological suitability of FPM. For this reason at least, the necessary evaluation of this important feature should be reassessed more times within the real shelf life of FPM because these materials are certainly exposed to chemical alterations [10] with the occurrence of microscopic and macroscopic phenomena also. At present, more research is certainly needed in relation to this problem.

1.5 The Predictable Behaviour of Food Packaging Materials in 'Normal Conditions'. Practical Applications

As mentioned above, the IFP can show different performances or predictable behaviours depending on several variables: edible raw materials, FPM, processing systems, storage conditions, etc. The same approach can be proposed when two FPM have to be evaluated. Sometimes, the simple comparison of the resulting IFP from two different processing lines (the same food or beverage and processing equipment, but different FPM) can be very helpful and economically interesting.

Actually, there are a number of different possibilities—testing methods, different protocols or conditions, etc. This matter is constantly evolving. Two different examples may be done here in relation to macroscopic failures.

The first situation is related to the so-called meshing effect [8]: the penetration of pigments and acid substances into the plastic coating of certain metal cans for pasteurised or sterilised sauces. Naturally, the effect is evident only on the inner side of these containers because of the following reasons:

(a) The acid nature of the packed food (tomato sauce, other pigmented vegetable products);
(b) The conditioning and packing process (hot temperature);
(c) The plastic nature of industrial enamels for metal cans (usually, white-coloured products).

The meshing effect is the appearance of little but macroscopic pinpoints on the white surface of these metal cans [8]. By the chemical viewpoint, it is known that red pigments and other acid substances may be transported under hot temperatures from the original sauce to the inner layers of the white enamel: this coating is substantially a sort of tridimensional matrix of organic polymers (usually, epoxyphenolic resins) with the presence of dispersed metal oxides (example: titanium dioxide). Organic pigments (carotinoids) may diffuse and place themselves into remaining matrix vacancies because of their chemical similarity with the organic structure. This effect is interesting because of two features:

1. Hot temperatures are needed, and the performance of white enamels can be evaluated in these conditions [17]
2. The appearance of diffused red points is generally permanent.

As a result, the food technologist (and the food auditor) should consider the possible risk caused by the evident diffusion of organic pigments from foods to FPM and vice versa. In other words, could the transfer of organic molecules be demonstrated in both directions? Apparently, the answer is positive: with exclusive reference to epoxyphenolic enamels, the presence and diffusion of different intermediates like bisphenol A and BADGE is well known and should counterbalance possible and visible transfers of red pigments from foods. Substantially, the diffusion is supposed to be a two-way process; the molecular dimension of red to yellow carotinoids and various acids should allow the concomitant displacement of plastic intermediates. More recently, the meshing effect has been discussed with reference to North African canned foods [18].

The second situation concerns the 'ghosting effect' in metal cans [1]. This phenomenon might be confused with the above-discussed set-off but the macroscopic detection is peculiar. In detail, the simple transfer by contact of printing inks from the external side to the inner surface of unfinished metal can bodies may be noted in several situations [10] with 'grotesque' effects (the appearance of strange printed images after sterilisation). Another similar situation is related to the recent detection of 2-isopropyl thioxantone (ITX, chemical formula: $C_{16}H_{14}OS$, CAS Number: 5495-84-1) in milk for babies; this time, used FPM were polycoupled containers and ITX was a common photoinitiator for ultraviolet printing inks [2].

Naturally, the possible danger is clear enough by the viewpoint of official authorities. On the other hand, it should be noted that above-discussed situations—meshing and ghosting effects—are both macroscopic failures and can be easily detected by official auditors and FP also. It can be easily concluded that the visual observation is simple enough and should be considered for the preliminary evaluation of the technological suitability and the assessment of chemical hazards in IFP [1].

References

1. Parisi S (2012) Food packaging and food alterations: the user-oriented approach. Smithers Rapra Technology, Shawbury
2. Piergiovanni L, Limbo S (2010) Materiali, tecnologie e qualità degli alimenti. Springer, Milan
3. Parisi S (2005) New implications of packaging in food products. Food Packag Bull 14(8 & 9):2–5
4. Italian Institute of Packaging (2009) Aspetti analitici a dimostrazione della conformità del food packaging: linee guida. Prove, Calcoli, Modellazione e altre argomentazioni. The Italian Institute of Packaging, Milan
5. Parisi S (2011) Food packaging and technological compliance. The importance of correct storage procedures. Food Packag Bull 20(9 & 10):14–18
6. British Standards Institution (2011) PAS 223:2011. Prerequisite programmes and design requirements for food safety in the manufacture and provision of food packaging. The British Standards Institution, London

7. Codex Alimentarius Commission (2001) Recommended international code of practice—general principles of food hygiene. The FAO/WHO food standards programme. http://www.fao.org/DOCREP/005/Y1579E/Y1579E00.HTM. Accessed 10 Oct 2013
8. Parisi S (2004) Alterazioni in imballaggi metallici termicamente processati. Gulotta Press, Palermo
9. Stilo A, Parisi S, Delia S, Anastasi F, Bruno G, Laganà P (2009) La Sicurezza Alimentare in Europa: confronto tra il 'Pacchetto Igiene' e gli Standard British Retail Consortium (BRC) ed international food standard (IFS). Ann Ig 21(4):387–401
10. Parisi S (2013) Food industry and packaging materials—performance-oriented guidelines for users. Smithers Rapra Technology, Shawbury
11. German Federation of Food Law and Food Science (2008) The 'declaration of compliance' for food contact materials and articles according to the German commodity ordinance. http://www.qsd.ie/wp-content/uploads/2012/04/Declaration-of-Compliance.pdf. Accessed 11 Oct 2013
12. German Federation of Food Law and Food Science (2012) The 'declaration of compliance' for plastic materials and articles intended to come into contact with food according to commission regulation (EU) No 10/2011 (plastics implementation measure, PIM). (BLL). http://www.bll.de/download/themen/bedarfsgegenstaende/konformitaetserklaerung-englisch-2012/. Accessed 11 Oct 2013
13. Vollmer A, Biedermann M, Grundböck F, Ingenhoff J-E, Biedermann-Brem S, Altkofer W, Grob K (2011) Migration of mineral oil from printed paperboard into dry foods: survey of the German market. Eur Food Res Technol 232:175–182. doi:10.1007/s00217-010-1376-6
14. Kernoghan N (2012) Mineral oil in recycled paper and board packaging. Smithers Pira. https://www.smitherspira.com/testing/food-contact/news-free-webinar-mineral-oil-in-recycled-paper-and-board-packaging.aspx. Accessed 11 Oct 2013
15. Coulier L, Bradley EL, Bas RC, Verhoeckx KC, Driffield M, Harmer N, Castle L (2010) Analysis of reaction products of food contaminants and ingredients: bisphenol a diglycidyl ether (BADGE) in canned foods. J Agric Food Chem 58:4873–4882. doi:10.1021/jf904160a. ISSN:0021-8561
16. European Food Safety Authority (2004) Opinion of the scientific panel on food additives, flavourings, processing aids and materials in contact with food (AFC) on a request from the commission related to the use of epoxidised soybean oil in food contact materials (Question N° EFSA-Q-2003-073) adopted on 26 May 2004 by written procedure. EFSA J 64:1–17. doi:10.2903/j.efsa.2004.64
17. Pilley KP (1981) Lacquers, varnishes and coatings for food and drink cans and for the decorating industry. Arthur Holden Surface Coatings Ltd., Birmingham
18. Parisi S, Laganà P, Gioffrè ME, Minutoli E, Delia S (2013) Problematiche emergenti di sicurezza alimentare. Prodotti etnici ed autenticità. In: Abstracts of the XXIV congresso interregionale siculo-calabro SitI, Palermo, 21–23 June 2013. Euno Edizioni, Leonforte, p 35

Chapter 2
Inorganic Contaminants of Food as a Function of Packaging Features

Angela Montanari

Abstract Metals are the most abundant group of chemical elements on the earth's crust and can be found in all foods. Some of them are essential to the diet, within certain specific tolerances, while others are present as contaminants and pose a risk to the human health. The knowledge of the risk by metal contamination in foodstuffs is an argument of great importance. Along the production chain, foods may come in contact with metals at different stages of the production process: parts of industrial plants, storage tanks, tools and mainly primary packaging. Some packaging materials are metallic; in other situations (plastics, etc.), metals are only one of components with a specific role. After an introduction on the international legislation, this chapter examines the main types of food containers—from metallic to plastic ones—considering the function of the metal, both as structural material or additive. For each material and packaging, factors affecting the related risk of contamination are analysed. Some case studies are examined referring to stainless steel, tinplate, aluminium, plastics and innovative packaging. The chapter concludes with a critical review with relation to some examples of metal concentration found in preserved foods, with a particular focus on heavy metals.

Keywords Corrosion · Engineered nanomaterial · European food safety authority · European regulation · Metal contamination · Migration · Specific migration limit

Abbreviations

Al	Aluminium
As	Arsenic
b.w.	Body weight
Cd	Cadmium
Ca	Calcium
CDC	Centers for Disease Control and Prevention
Cr	Chromium
Co	Cobalt

© The Author(s) 2015   17
C. Barone et al., *Food Packaging Hygiene*, Chemistry of Foods,
DOI 10.1007/978-3-319-14827-4_2

Cu	Copper
ECCS	Electro-coated chromium steel
ENM	Engineered nanomaterial
EDI	Estimated daily intake
EFSA	European Food Safety Authority
EU	European Union
FAO	Food and Agriculture Organization
FACET	Flavourings, Additives and Food Contact materials Exposure Task
FCM	Food contact material
ICP-MS	Inductively coupled plasma—mass spectrometry
TOF-ICP-MS	Inductively coupled plasma-time of flight-mass spectrometry
Fe	Iron
JECFA	Joint FAO/WHO Expert Committee on Food Additives
Pb	Lead
LoQ	Limit of quantification
Li	Lithium
Mg	Magnesium
Hg	Mercury
DM	Ministerial Decree
Ni	Nickel
AFC	Panel on Food Additives, Flavourings, Processing Aids and Food Contact Materials
ppb	Part per billion
ICP-AES	Plasma atomic emission inductively coupled spectroscopy
PP-g-PAA	Polypropylene-grafted-poly(acrylic acid)
PE	Polyethylene
PTWI	Provisional tolerable weekly intake
RASFF	Rapid Alert System for Food and Feed
SML	Specific migration limit
SSICA	Stazione Sperimentale per l'Industria delle Conserve Alimentari
THQ	Target hazard quotient
Sn	Tin
TFS	Tin-free steel
Ti	Titanium
V	Vanadium
Zn	Zinc
WHO	World Health Organization

2.1 Introduction

Metals are the most abundant group of chemical elements on the earth's crust, and they are found in all foods. Some of these elements, such as iron, calcium, potassium and zinc, are present in nature and are considered essential when speaking of human diet at least, within certain specific tolerances. On the other hand, metals, such as lead, cadmium, arsenic and mercury, may be detected in foods and other commodities as contaminants and pose serious risks to the human health because of different factors, including the known bioaccumulation. Table 2.1 shows main effects on the human health resulting from deficiency or excess of certain metals.

The knowledge of the contribution of certain metals in various food matrices is extremely important for different reasons, including nutritional purposes and the necessity of preventing contamination episodes by toxic metals. By a general viewpoint, the detection of metals in preserved foods can have three main causes:

- Presence in raw materials used in the preparation of preserved foods. Metallic elements may be naturally present in raw materials. On the other hand, the detection of metals may depend on environmental contamination
- Presence in food preparations before of the final packaging. The cause(s) can be originated on one or more of processing steps. Examples: contact with metal parts of processing plant (tubes, thanks, valves and electrodes)
- Contamination of preserved foods during packing and especially during storage steps.

Depending on the level of contamination, several corrective actions have to be put in place including (a) analyses of raw materials, (b) evaluation of production steps and (c) the examination of packaging and/or distribution processes.

Table 2.1 Main adverse health effects of certain metals

Metal	Deficiency	Surplus
Calcium	Bone deformities; osteoporosis	Cataract; stones cock; arteriosclerosis
Chromium	Glucose's metabolic disorders	Lung cancer
Cobalt	Anaemia	Heart problems
Iron	Anaemia; kinky hair syndrome (Menke's)	Cirrhosis; neuropathies; Wilson's disease
Cuprum	Anaemia	Primary and secondary haemochromatosis; haemosiderosis; cirrhosis
Litium	Depression	
Magnesium	Nervous disorders; weakness; stunted growth	Anaesthetic
Manganese	Skeletal deformities; gonads dysfunctions	
Kallium	Muscle cramps; muscle weakness; paralysis	Addison's disease
Selenium	Liver necrosis	Fluid restriction; high blood pressure
Sodium	Addison's disease; lack of appetite; apathy; muscle cramps	

2.2 Legislation

The presence of metals in foods is regulated through two series of laws relating to the final product on the one part and to packaging materials (containers) on the other side.

With reference to preserved and packaged food products, the Regulation (CE) No. 1881/2006 and subsequent updates, lastly the Reg. (UE) No. 420/2011 [1] and (UE) 488/2014 [2], set limits on the content of different toxic metals: lead (Pb), cadmium (Cd), mercury (Hg) and tin (Sn) in foods. In addition, the Reg. (CE) No. 333/2007 [3], modified from the Reg. (UE) 836/2011 [4], defines methods of sampling and analysis for the official control of Pb, Cd, Hg, inorganic Sn, 3-monochloropropane-1,2-diol and polycyclic aromatic hydrocarbons in foods.

With relation to food packaging materials, several European and national rules govern packaging and materials in contact with food. Actually, the matter of food packaging legislation in the European Union (EU) is extremely complex. Normally, the EU legislation on food packaging can be subdivided in two different groups:

- General rules, which concern all the materials. These norms define fundamental requirements for a food contact material or object
- Specific rules, with relation to individual materials. There are only some specific rules at the European level: the main of these legislations concerns substantially plastic materials, while other legislative documents are directly correlated with the control of ceramic materials and cellulose.

In general, three fundamental points have to be mainly considered as the pilasters of these rules:

- Composition requirements: compliance with the so-called positive lists
- Specific migration limits (SML)
- Prohibited materials.

The 'General Framework Regulation' for all FCM is the Regulation (EC) No. 1935/2004 [5].

On the other hand, it has to be observed that specific requirements for metals and alloys used in food contact materials and articles are not defined at present in the EU legislation.

In detail, the following EU Member States have specific legal provisions or official recommendations on metals for food contact applications: Austria, Finland, France, Germany, Greece, Netherlands, Norway and Sweden. These provisions cover mainly the transfer of heavy metals from metallic food contact articles into foodstuff. Italy is definitely the country with the largest number of specific regulations for individual materials. For this reason, the main Italian legislation for food packaging, the Ministerial Decree (DM) 21 March 1973 and subsequent updates (DM 18 April 2007 no. 76 on aluminium and DM 21 December 2010, no. 258 on stainless, now repealed by the D.M.11 November 2013, no. 140) is often cited in this text [6–8].

Table 2.2   Use of metals in the modern industry of food contact materials

Main function or industrial uses	Main applications
Structural material	Tinplate 'Tin-free Steel' (TFS) or 'Electro-coated Chromium Steel' (ECCS) Aluminium
Additives and processing aids	Fillers Stabilisers Dyes
Active packaging	Oxygen scavengers Gas barrier agents Antimicrobic agents Antioxydants
Nanomaterials	Nanocompounds (Ag, Ti, Zn)

This series of rules regulates the use of metals (Table 2.2) when used as the main and structurally basis of containers (tinplate cans are one of the main examples) or considered as additives for packaging materials and objects (fillers and pigments in plastics). There are not harmonised documents with relation to the use of stainless steel at present.

Anyway, Article 3 of the Regulation (EC) No. 1935/2004 is considered and applied when speaking of specific non-regulated materials [5]. In detail, Article 3 clearly states that materials and article for food contact applications are not allowed, under normal or foreseeable conditions of use, to transfer their constituents to food in quantities which could:

- Damage the human health
- Modify the composition of the packaged food in an unacceptable way
- Cause the deterioration of sensorial features of the packaged food.

Recently, a new Resolution on metals and alloys used in food contact materials and articles has been published in December 2013 with the aim of overcoming the lack of specific regulations materials in the EU [9].

With specific relation to health risks arising from consumer exposure to certain metal ions, the above-mentioned Resolution recommends the adoption of legislative actions and other measures to the Member States. Substantially, health hazards are defined with relation to the detection of metal ions when released to food from food contact metals and alloys during manufacture, storage, distribution and use. The Resolution provides detailed principles and guidelines in the annexed Technical Guide on Metals and Alloys used in food contact materials and articles (first edition). Above-mentioned documents have been prepared in cooperation with European experts in this field from national authorities, manufacturers and private testing laboratories.

Interestingly, this Resolution defines quality requirements for materials such as aluminium foil, kitchen utensils and coffee machines without specific EU limits. For example, the release of nickel should not exceed 0.14 mg/kg, while Pb should not be released in amounts greater than 0.0043 mg/kg (this amount is intended as the detected concentration of metal ions in food).

In addition, detailed instructions on laboratory testing are described in the Guideline [9], with specific relation to analytical methods for migration testing of food contact materials and articles made from metals and alloy. Finally, the technical Guideline provides necessary advices with concern to the preparation of the Declaration of Compliance (Sect. 1.3) for metals and alloys used in food contact materials and articles. In detail, the list of structural metals includes the following elements:

- Aluminium (Al)
- Antimony
- Chromium (Cr)
- Cobalt (Co)
- Copper (Cu)
- Iron (Fe)
- Magnesium (Mg)
- Manganese
- Molybdenum
- Nickel (Ni)
- Silver
- Sn
- Titanium (Ti)
- Vanadium (V)
- Zinc (Zn).

It has to be also considered that other metal contaminants and impurities can be examined: this group includes arsenic, barium, beryllium (Be), Cd, Pb, lithium (Li), Hg, thallium, stainless steel and other alloys.

2.3 Analytical Methods

At present, official methods for the analysis of metal contaminants (traces) in food matrices are given as follows:

- Flame and graphite furnace atomic absorption spectroscopy and
- Plasma atomic emission inductively coupled spectroscopy (ICP-AES).

The ICP-AES ensures an excellent analytical sensitivity when coupled with a mass spectrometer. This system, the 'Inductively Coupled Plasma-Mass Spectrometry' (ICP-MS), can determine metal concentrations below 10 parts per billion (ppb). In addition, the new 'Inductively Coupled Plasma-Time Of Flight-Mass Spectrometry' (TOF-ICP-MS) system (Argon plasma) enables faster analyses and allows higher precision in the isotopic analysis.

Moreover, electro-analytical techniques have been developed considerably in recent years. These methods can guarantee high precision and analytical sensitivity; the small size of necessary instruments will favour the transport even for in situ analysis.

Normally, analyses are carried out using spectroscopic techniques after the proper preparation of samples by treatment with acids. Recently, several works have proposed new methods for sample preparation and analysis with increased sensitivity and the reliable determination of trace toxic contaminants.

For example, the development of sensitive and reliable analytical techniques for the precise monitoring of lead in various foodstuffs has been reported [10]. In detail, the enrichment and separation procedure for lead has been proposed prior to its flame atomic absorption spectrometric determination [10]. In these conditions, a very low limit of detection of Pb has been reported: 0.36 μg/l (3σ, $n = 7$). According to researchers, the application of this method to the determination of trace lead in beer and tea drinks may be proposed [10].

Other techniques have been recently developed and validated with concern to the determination of As, Cd and Pb contents by means of quadrupole ICP-MS [11]. These metals, of big concern when speaking of the human health, can easily enter the food chain through the environment and/or as a consequence of food manufacturing processes. As a result, foodstuffs may be considered the main human exposure route to these chemical elements [11]. For these reasons, the European Food Safety Authority (EFSA) recommends the reduction of the exposure to Cd and Pb so as to protect especially vulnerable subgroups of population (e.g. infants). On this basis, the availability of precise, accurate and sensitive analytical methods for the reliable detection of low concentration values is a key point especially for official control laboratories.

According to researchers, the determination of As, Cd and Pb contents by means of quadrupole ICP-MS can allow following limit of quantification (LoQ) values: 6.2, 1.2 and 4.5 μg/kg for As, Cd and Pb, respectively, in strict accordance with requirements set in the Commission Regulation (EC) No. 333/2007 [11]. Pb and Co contamination in tap water and food samples can be also detected with a new procedure based on the formation of complexes of metal ions with 8-hydroxyquinolein in aqueous solution [12]. According to researchers, the preconcentration and separation of metals by solid-phase extraction (with paper filter) can be followed by spectrofluorimetric determination. Detection limits have been found to be 0.043 and 0.0219 μg/l (signal/noise = 3) for Pb(II) and Co(II) ions, respectively, [12]. The new methodology has obtained satisfactory results when speaking of the determination of trace amounts of Pb and Co in foods samples (milk powder, express coffee and cocoa powder) and tap waters from different regions of Argentina [12].

2.4 Metal Contamination and Toxicology

The toxicological risk evaluation, also in the case of metals, is based on two key factors in different situations, including also metallic contaminants: (a) the hazards of the migrating substance and (b) the correlated amount. Different factors have to be taken into account: the nature and the composition of the material, the type and the composition of surfaces, the temperature and time of contact. In addition, the evaluation of the exposition of every single metal is crucial.

The 'Flavourings, Additives and Food Contact materials Exposure Task' (FACET) European project has given an important contribution in terms of the creation of a database containing information on levels of different food-related substances and corresponding food consumption data [13]. The covered packaging materials have been plastics (flexible and rigid materials), metal containers, light metal packaging, paper and board materials, as well as used adhesives and inks. This project has established a migration modelling framework for packaging materials into foods under real conditions of use. On these bases, the realistic estimation of substance concentrations for consumer exposure modelling has been obtained with the consequent creation of a reliable food intake database [13]. Generated data can provide exposure estimations using probabilistic models. It has to be also noted that the evaluation of exposure is expressed for individual consumers and various percentiles of different populations and subpopulations, when covered by national dietary surveys.

2.4.1 Aluminium

At present, there is no indication of any adverse health effects caused by released aluminium from packaging material, when speaking of packaged food products.

The Joint FAO/WHO Expert Committee on Food Additives (JEFCA) of the Food and Agriculture Organization (FAO) and the World Health Organization (WHO) has established a 'Provisional Tolerable Weekly Intake' (PTWI) of 1 mg/kg body weight (b.w.) for aluminium in 2006 [14]. This limit applies to all aluminium compounds in food, including additives. The EFSA has adopted the same PTWI in 2008 [15]. Subsequently, the European Commission (EC) has reviewed use levels and conditions of use for aluminium-containing food additives. The Commission Regulation (EU) No. 380/2012 [16] amends several provisions in Annex II to the Regulation (EC) No. 1333/2008 relating to aluminium and aluminium lakes; Annex II contains a positive list of additives approved for use in food in the EU and their permitted conditions of use.

2.4.2 Tin

At present, there is no indication of a chronic toxicity of Sn in humans because this element does not accumulate in the organism (traces in the bones > soft tissues). The acute toxicity of Sn is rather low: according to a recently published study, tin levels up to 267 mg/kg in foodstuff do not cause any harm to the health of adults. It should be noted that there is a great variation in the sensitivity of individuals to Sn. Different levels for chronic and acute toxicity of Sn could be established.

2.4.3 Lead

The human exposure to Pb causes a variety of health effects with particular relation to children. People are exposed to Pb through the air they breathe, through water and through food/ingestion. Toxic effects are usually due to long-term exposure. The Centers for Disease Control and Prevention (CDC) in the United States of America has defined 10 µg/dl of whole blood as the reference blood Pb level for adults [17]. This level is reduced when speaking of children: 5 µg/dl of blood [17]. On the other hand, the maximum limit for Pb in canned tomato paste is 1.0 mg/kg according to the Codex Standard 193-1995 [18].

2.4.4 Cadmium

Oral exposure to Cd may determine adverse effects on a number of human tissues, including also the immune system, and the cardiovascular system [19]. The intake of Cd from the diet is usually about 0.0004 mg/kg/day, roughly ten times lower than the typical amount needed to cause kidney damage by this route. With reference to this metal, the Codex Alimentarius Commission has defined a limit of 0.05 mg/kg [18].

2.4.5 Arsenicum

Inorganic As is well known as a notable human carcinogen; in addition, children can suffer other health problems in later life. Available data have shown that inorganic As causes cancer of the lung and urinary bladder, in addition to skin damages. There are no limits for As in most foods with relation to the USA, but the recognised standard value for drinking water is 10 ppb. With concern to the European viewpoint, the EFSA has recommended that the dietary exposure to inorganic As should be lowered in comparison with the JECFA PTWI of 15 µg/kg b.w. [20].

2.4.6 Rapid Alert System for Food and Feed

The current situation of food contamination in the EU can be reliably monitored by means of the 'Rapid Alert System for Food and Feed' (RASFF). This tool can be useful because of the possibility of exchanging rapidly information on measures taken to ensure food safety among the member States of the EU. Food contact materials (FCM) are included in the group of categorised products of interest for the RASFF. As an example, 516 episodes of alert have been notified in 2012 with relation to Italy only: 95 of these notifications have concerned FCM. By a general viewpoint, main causes of rejection appear to be the release of heavy metals and a high level of total migration.

2.5 Packaging Materials: Examples of Applications

2.5.1 Stainless Steel and Glass

Both stainless steels and glass for FCM production contain heavy metals, Ni, Cd and Pb: all these elements can migrate into foods. With relation to metal contamination in foods and the correlated risk management, the Italian legislation can be taken as a reference. In particular, the DM 21/03/73 [6] and subsequent updates include both regulatory compliance of the composition and SML. With exclusive reference to Italian norms, available maximum limits are 0.1 mg/kg for Cr and Ni, and 0.3 mg/kg for Pb. The migration of these metals is easily observed. For example, Table 2.3 shows several concentration values of these metals in tomato puree and lemon juice after 12 months of storage under nitrogen at room temperature: these determinations have been carried out by the Italian *Stazione Sperimentale per l'Industria delle Conserve Alimentari* (SSICA). Sampled foods have been stored in tanks manufactured in three different materials: stainless steel AISI 304, stainless steel AISI 316 and titanium. Results have shown that the overcoming of limit values for different metals is mainly dependent on the aggressiveness of the product, while the type of packaging material does not appear to have a similar influence. AISI 304 tanks appear to show the lower resistance to corrosion phenomena, in agreement with the literature and productive experiences.

2.5.2 Metallic Cans and Tubes

At present, the limit of migration for Sn is defined by European regulations (EC) No. 242/2004 [21] and (EC) No. 1881/2006 [1] when speaking of tinplate cans. In detail, Sn limits are established as a function of the kind of foodstuffs—from 50 to 200 mg/kg—as shown in Table 2.4. According to the EN 10333/2005 norm, the use of Sn is also foreseen with a minimum degree of purity of 99.85 % with the aim of reducing the content of heavy metals such as Pb [22] (maximum allowed concentration in tin coatings have to be lower than 0.01 %).

Moreover, the Italian DM 18 February 1984 sets a limit of 50 mg/kg for Fe in food products, while Pb is allowed between 0.2 and 3.0 mg/kg [23].

The corrosion process of tinplate cans is very complex, depending on a large number of parameters. Briefly, it can be observed that corrosive phenomena occur mainly on tin coating in internally plain containers. On the other side, the risk of finding high concentrations of Fe in canned foods is greater when cans are internally lacquered.

This behaviour of tinplate cans is exemplified in Table 2.5 and in Fig. 2.1. In particular, Fig. 2.1 shows that Sn limits are not exceeded even in the most unfavourable thermal condition.

Table 2.3 Detection of heavy metals in tomato puree and lemon juice after 12 months of storage under nitrogen at room temperature

Metal concentration (mg/kg)	Packaged food product (storage: 12 months of storage under nitrogen at room temperature)							
	Tomato puree (*passata*)				Lemon juice			
	Unpackaged	AISI 304	AISI 316	Titanium	Unpackaged	AISI 304	AISI 316	Titanium
Iron	3.32	3.78	3.70	2.94	0.70	65.3	5.42	2.66
Chromium	0.03	0.10	0.10	0.06	0.05	7.55	0.50	0.20
Nickel	<0.01	0.14	0.16	0.10	<0.01	5.50	0.16	0.14
Molybdenum	<0.02	<0.02	<0.02	<0.02	<0.02	<0.02	<0.02	<0.02
Titanium	<0.01	<0.01	<0.01	<0.01	<0.01	<0.01	<0.01	<0.01

These foods have been stored in different tanks: the difference concerns the structural material of containers (stainless steel AISI 304, stainless steel AISI 316 and titanium). Apparently, the overcoming of limit values for different metals is mainly dependent on the aggressiveness of the product and on the type of material. AISI 304 tanks appear to show the lower resistance to corrosion phenomena

Table 2.4 Maximum allowed limits for inorganic tin, according to the Reg. (EC) No. 242/2004, Annex I, Sect. 6 [21]

Product	Maximum level (mg/kg wet weight)	Performance criteria for sampling	Performance criteria for methods of analysis
1. Canned foods other than beverages	200	Commission Directive 2004/16/EC	Commission Directive 2004/16/EC
2. Canned beverages, including fruit juices and vegetable juices	100	Commission Directive 2004/16/EC	Commission Directive 2004/16/EC
3. Canned foods for infants and young children, excluding dried and powdered products	50	Commission Directive 2004/16/EC	Commission Directive 2004/16/EC
3.1. Canned baby foods and processed cereal-based foods for infants and young children (1)	50	As above mentioned	As above mentioned
3.2. Canned infant formulae and follow-on formulae, including infant milk and follow-on milk (2)	50	As above mentioned	As above mentioned
3.3. Canned dietary foods for special medical purposes (3) intended specifically for infants	50	As above mentioned	As above mentioned

(1) Baby foods and processed cereal-based foods for infants and young children as defined in Article 1 of Directive 96/5/EC. Maximum level refers to the product as sold
(2) Infant formulae and follow-on formulae as defined in Article 1 of Directive 91/321/EEC. Maximum level refers to the product as sold
(3) Dietary foods for special medical purposes as defined in Article 1(2) of Commission Directive 1999/21/EC of 25 March 1999. Maximum level refers to the product as sold

Table 2.5 Tin contamination in peeled tomatoes

Capacity (kg)	Sn storage at 20 °C (mg/kg)	Sn storage at 37 °C (mg/kg)
0.5	112	147
1.0	122	132
3.0	92	121

Concentrations of Sn in plain cans of different capacity filled with peeled tomatoes after 9 months of storage (two different temperatures)

Fig. 2.1 Influence of storage temperatures on iron concentration in canned food products at different times. Metal cans are internally protected with white enamels

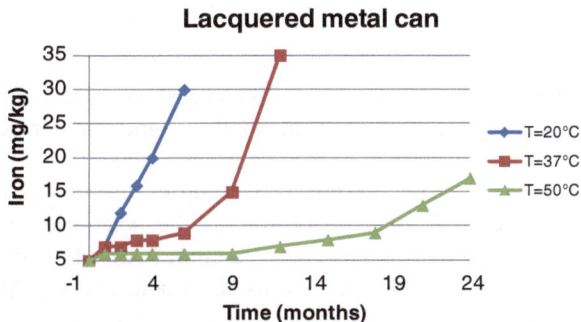

Table 2.6 Release of Al in canned tea products (different storage times, plain and lacquered aluminium cans)

Release of Al in different storage conditions	Plain sample	Lacquered sample
After 7 days, mg	3.25	
After 26 days, mg	7.69	
In steady-state conditions between 7 and 26 days		
mg	4.44	
mg/dm^2/day	1.17	
After 51 days		
mg		0.31
mg/dm^2		1.55

With relation to 'Tin-Free Steel' (TFS) or 'Electro-coated Chromium Steel' (ECCS) cans, the Italian reference is the DM No. 243 of 01st June 1988 [24]. According to this document, maximum allowed values for Cr in canned foods are 0.4 mg/kg (in four of the five samples) and 0.5 mg/kg (in the remaining sample). On the other hand, Fe cannot exceed 50 mg/kg.

With concern to Al, the Italian DM No. 76 of 18th April 2007 does not apply to aluminium materials and articles when coated with an organic film, such as cans and tubes [24]. On the other side, this legislation applies to all other uses of Al (e.g. foils and trays) and determines the degree of purity of Al and the composition of its alloys. At the European level, the following standards apply in relation to the chemical composition: EN 601, EN 602, EN 14287 and EN 13046/2000 norms [25–28]. Both materials, TFS and aluminium, are always protected with a lacquer when used for preserved foods; consequently, the migration of Cr and Al is reduced and there are not recognised limits. As an example, stringent corrosion tests (SSICA researches) on lacquered TFS samples in citric acid solutions (pH = 4) have shown the following values of Cr migration after 1 month of storage at 37 °C: 13.9–22.6–37.5–31.2–14.5 µg/kg. Anyway, the legally allowed maximum value of 400 µg/kg has not been exceeded. Migration values appear low in other situations concerning two pieces of aluminium cans for beverages with an organic coating (Table 2.6).

2.5.3 Regenerated Cellulose Films

The Commission Directive 2007/42/EC regulates materials and objects of regenerated cellulose film when intended to come into contact with foods [29].

With concern to these materials, metals have the function of additives: the quantity of each substance or group of substances must not exceed 2 mg/dm^2 of the uncoated film. According to the scientific literature, different molecules—oxides and hydroxides of Al, calcium (Ca), Mg and silicon; silicates and hydrated silicates of Al and Ca—are used. In addition, Ca, Mg, potassium and sodium are detectable because of the presence of related salts. As an example, Zn and Cr may be found up to 50–70 and 56–15 mg/kg, respectively, in different packaged foods after contact with recycled paper [7].

2.5.4 Plastic Materials

At the European level, the fundamental legislation for plastic materials is the Commission Regulation (EU) No. 10/2011: it determines SML or maximum usable amounts for each metal or metal salt included in the list [30]. In fact, metals are quite frequently used as components of plastic materials with different functions: dyes, fillers, pigments, antifouling, gas barrier agents, etc.

When speaking of plastic materials and metal components, a premise should be done because of the necessity of distinguishing between paints or enamels, adhesives or compound and plastic films.

First of all, titanium dioxide, zinc oxide or carbonate, aluminium oxide and barium sulphate are mainly used in paints as pigments. Titanium dioxide, barium sulphate, calcium carbonate and magnesium silicate or silicates of calcium and magnesium are used as inorganic fillers for the production of compounds for caps and cans.

Plastic films may easily contain similar substances: titanium dioxide, calcium carbonate and magnesium, silicates of calcium and magnesium, salts of cadmium, molybdenum, chromium, copper, gold and silver. Some of these additives—aluminium oxide, cobalt oxide, manganese oxide, calcium butyrate, calcium chloride, calcium hydroxide and calcium oxide—can be used without limits. On the opposite hand, the addition of certain metals may be restricted: for example, SML for antimony trioxide cannot exceed 0.04 mg/kg as antimony.

With relation to the Commission Regulation (EU) No. 10/2011, Annex II concerns the following SML restrictions for metals (analytical values are referred to foods or food simulants):

- Barium: 1 mg/kg
- Cobalt: 0.05 mg/kg
- Copper: 5 mg/kg
- Iron: 48 mg/kg
- Lithium: 0.6 mg/kg

- Manganese: 0.6 mg/kg
- Zinc: 25 mg/kg.

A useful and fast analytical technique to identify metals in a plastic film, compound or lacquer is the electronic scanning microscopy coupled to X-ray microanalysis. Figure 2.3 shows an example of application with relation to the analysis of a compound for caps (SSICA researches). The presence of barium sulphate has been detected in the area as shown in Fig. 2.2 (300× magnifications) as indicated by the X-ray spectrum (Fig. 2.3).

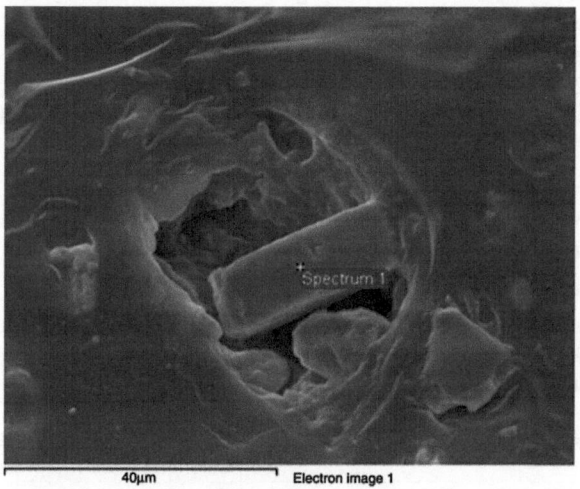

Fig. 2.2 Superficial deposition of barium sulphate on food contact caps (300× magnification). The detection of this sulphate has been confirmed by means of electronic scanning microscopy coupled to X-ray microanalysis (Fig. 2.3)

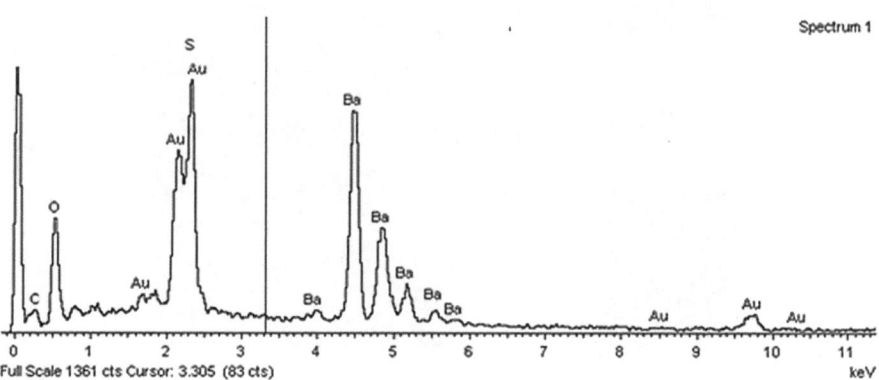

Fig. 2.3 Analytical detection (X-ray spectrum) of barium sulphate in a compound for food contact caps (Fig. 2.2). The analytical procedure concerns the use of electronic scanning microscopy coupled to X-ray microanalysis

2.5.5 Active and Intelligent Packaging: Nanotechnologies

The development of active packaging and the correlated use has progressively grown up in recent years: metals are used in this field with many functions, as briefly shown in Table 2.7. Active packaging instruments are ruled by the Regulation (EC) No. 450/2009 [31]. One of these devices, used in the packaging of sliced cooked hams, is shown in Fig. 2.4: the picture shows the section of a polyethylene (PE) film added with iron particles as oxygen scavenger (SSICA researches).

Actually, the sector of active packaging devices is notably diversified and in continuous evolution. Some examples can be reported here.

For instance, an innovative used of metals in plastic material has been recently discussed in 2014 [32]. Researchers have developed formulations of low-cost bio-based oxo-biodegradable PE/lignin hybrid polymeric composites prepared by

Table 2.7 Active packaging systems. A brief classification of main categories and correlated 'active' components

Active packaging categories	Chemical components
O$_2$ scavengers	Clays
Humidity absorbers	Clays
Humidity regulators	Potassium chloride Sodium chloride
Carbon dioxide scavengers	Calcium chloride + sodium hydroxide Calcium chloride + potassium hydroxide
Antimicrobic agents	Titanium dioxide
Carbon dioxide emitters	Ferric carbonate + metal

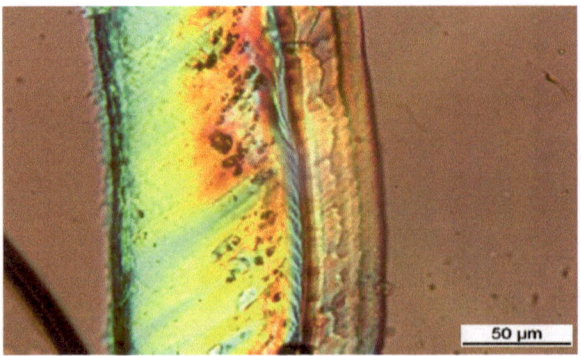

Fig. 2.4 This picture shows the image of a section of a polyethylene film added with iron particles as oxygen scavenger (food contact application: packaging of sliced cooked hams). The amount of iron-based oxygen scavenger is approximately 12 %

using ethylene vinyl acetate copolymer as compatibiliser and a transition metal salt as oxo-biodegradation promoter. The final aim has been the mitigation of the environmental burden caused by plastic waste items [32].

Another application [33] demonstrates the ability to tailor chelating activity of 'polypropylene-grafted-poly(acrylic acid)' (PP-g-PAA) with potential applications in active packaging. The development of iron chelating films prepared by photoinitiated graft polymerisation of acrylic acid on polypropylene can be very useful because Fe (and other transition metals) can enhance the oxidative degradation of lipids. In other words, active packaging labels with non-migratory metals can surely meet consumer demand for 'cleaner' labels. Substantially, the active PP-g-PAA-based material has been produced with a ligand (carboxylic acid)/metal (Fe^{2+}) binding ratio of $\sim$4–5 [33].

In addition, nanotechnologies are used in food packaging: metal nanoparticles are used as tools for improving gas barriers, in particular oxygen. Nanoparticles may also act as nanosensors. At present, there is still no specific legislation. However, the evaluation of possible risks to the human health and to the environment should be done: unmetabolised nanoparticles can cause serious and incurable diseases to the human being. Recently, the EFSA has published the 'Guidance on risk assessment concerning potential risks arising from applications of nanoscience and nanotechnologies to food and feed' [34]. This document aims to discuss the characterisation, exposure scenarios and hazard identification for 'engineered nanomaterials' (ENM).

Basically, the use of nanoparticles in plastic packaging must be authorised by the EFSA in accordance with the Reg. (EU) No. 10/2011, Art. 9 (2): 'Substances in nanoform shall only be used if explicitly authorised and mentioned in the specifications in Annex I' [30]. For example, a peculiar restriction concerns titanium nitride (Annex I), in accordance with Annex I and an EFSA Opinion in 2012 [35]: 'No migration of titanium nitride nanoparticles. Only to be used in PET bottles up to 20 mg/kg. In the PET, the agglomerates have a diameter of 100–500 nm consisting of primary titanium nitride nanoparticles; primary particles have a diameter of approximately 20 nm'.

Because of the growing importance of nanotechnologies and possible consequences on the human health, the 'Scientific Network for Risk Assessment of Nanotechnologies in Food and Feed' (Nano Network) has been launched in February 2011. As a result, this organisation is expected to give an annual report on 'Risk Assessment of Nanotechnologies in Food and Feed' [34].

With concern to last scientific literature reviews, it has been recently outlined that the use of nanotechnology-derived food could be connected with the potency to lead to systemic toxicity [36]. This conclusion has been reported on the basis of existing data on the (potential) use of ENM in the food industry, including available information on toxicity profiles of commonly applied ENM such as metal (oxide) nanoparticles. In addition, researchers have also highlighted major gaps that need further research and regulation in this field [36].

From the analytical viewpoint, recent papers appear to highlight the importance of the quantitative amount of added amounts. For example, a new analytical method based on ICP-MS has been developed with the aim of determining the migration of titanium from nano-titanium dioxide-PE films used for food packaging into food simulants under different temperature and migration time conditions [37]. In detail, researchers have found that the maximum migration amounts into 3 % (w/v) aqueous acetic acid were 12.1 ± 0.2 μg/kg at 100 °C (the highest thermal values). On the other hand, maximum migration values of Ti were 2.1 ± 0.1 μg/kg into 50 % (v/v) aqueous ethanol [37]. Briefly, researchers have revealed that the increase of additive contents in films may promote the migration of nanoparticles. In addition, nanoparticles appear to migrate via dissolution from the surface of films into the liquid phase (food simulant) [37].

2.6 Metals, Diet and Preserved Foods

With concern to the content of metals in preserved foods (restricted geographical areas or specific products), several papers are available at present. For example, a detailed study on the dietary exposure to several metals discusses available data with reference to the diet of adult British citizens in 1997 [38]. In detail, this research has demonstrated that the dietary exposure at the level of confidence of 97 % was 5.7, 0.024 and 1.9 mg/day per Al, Pb and Sn, respectively; in addition, detected results were below the official PTWI of 60, 0.21 and 120, respectively. It has been reported also [38] that the main sources of contamination were bread, cereals and fish (for Al), bread and nuts (for Pb) and finally tin canned vegetable products (for Sn).

Another work has concerned the evaluation of heavy metals contamination in Iranian canned tomato paste and tomato sauce (ketchup) [39] during the period 2010–2013. In summary, obtained results for Pb, Cd and As have been found lower than the limits of national and international standards in all samples. It has been reported that the average concentration of As was 62 ± 14 and 48 ± 12 ng g^{-1}, while Cd values were below the LoQ in 7 % of tomato paste and 10 % of ketchup samples. Finally, Pb concentrations have been estimated below the LoQ in 75 % of tomato paste and 77 % of ketchup samples [39].

Similar works have concerned Cd, Pb and other metals in peculiar products of the *Maghreb*. For example, levels of Cd, Pb and Hg have been detected in fish from the Atlantic sea (Morocco) by the Moroccan Reference Laboratory as part of a specific monitoring program in 2014 [40]. Obtained results have confirmed that contamination amounts in muscles of fish correspond to the following values: 0.009–0.036, 0.013–0.114 and 0.049–0.194 μg/g for Cd, Pb and Hg, respectively. As a consequence, researchers have concluded that fish and shellfish from southern areas of Morocco should not cause health problems for consumers [40]. Anyway, maximum residual levels have been found within the maximum residual levels prescribed by the EU.

With exclusive relation to fruits and vegetables consumed in Algeria, another paper has found that the estimated daily intake (EDI) and the target hazard

Table 2.8 Packaged foodstuffs in metallic containers and metal contamination: fish products

Fish products in lacquered metallic cans: metal contamination

Fish product	Average values (mg/kg)				Maximum allowed admitted (mg/kg)			
	Lead	Cadmium	Mercury	Tin	Lead	Cadmium	Mercury	Tin
Tuna in olive oil	0.02	0.02	0.15	<3.0	0.30	0.05	1.0	200
Mackerel filet in olive oil	<0.03	<0.03	<0.1	<3.0	0.30	0.05	1.0	200

Table 2.9 Packaged foodstuffs in metallic containers and metal contamination: meat products

Meat products in lacquered metallic cans: metal contamination

Meat product	Average values (mg/kg)				Maximum allowed concentration (mg/kg)			
	Lead	Cadmium	Mercury	Tin	Lead	Cadmium	Mercury	Tin
Meat of bovine	0.03	<0.03	(1)	<3.0	0.10	0.05	(2)	200
Turkey meat	0.04	<0.03	(1)	<3.0	0.10	0.05	(2)	200
Meat of chickens	0.03	<0.03	(1)	<3.0	0.10	0.05	(2)	200
Pork	0.04	<0.03	(1)	<3.0	0.10	0.05	(2)	200
Würstel	0.07	<0.03	(1)	<3.0	0.10	0.05	(2)	200
Medallions cattle	0.04	<0.03	(1)	<3.0	0.10	0.05	(2)	200

(1) This metal has not been researched in analysed samples
(2) There are not official maximum allowed concentration limits with concern to Hg in these products

quotient (THQ) may be defined below threshold values for Cu, Zn and Cr; on the other side, Pb values have been judged excessive (EDI: 15.66 μg/kg b.w/day; THQ: 4.37), indicating an obvious health risk over a lifetime of exposure [41].

With concern to wines, trace **metal** contents have been studied for the first time in Italy [42] over the period 1995–2010. In summary, researchers have found that the decreasing use of pesticides and phytoiatric products has progressively determined the decrease of Cd and Cu residues in wines. At the same time, a significant decrease (about 74 %) has been observed for Pb from 1995 to 2010 [42], probably because of the diminution of Pb emissions in the atmosphere following the phasing out of metal from gasoline (in Italy since 2002).

The Italian SSICA has performed numerous analyses of the content of heavy metals in preserved foods. Tables 2.8, 2.9, 2.10, 2.11, 2.12, 2.13 and 2.14 show analytical results per different typologies of food product. Average results have been obtained with relation to different packs of the same lot.

In summary, the amount of Hg has been found below the detection limit of the instrument (<0.01 mg/kg) on all below-listed products:

• Guar gum (E412) and agar agar (E406)
• Food emulsifiers for mayonnaise, yogurt, ice cream, meat

Table 2.10 Packaged foodstuffs in metallic containers and metal contamination: cereals and cereal-based products

Cereals and cereal-based products in lacquered metallic cans: metal contamination

Food product	Average values (mg/kg)			Maximum allowed concentration (mg/kg)		
	Lead	Cadmium	Tin	Lead	Cadmium	Tin
Pasta	0.03	<0.03	<3.0	0.2	0.02	200
Rice salad	0.03	<0.03	<3.0	0.2	0.02	200
Vegetable soup	0.04	<0.03	<3.0	0.2	0.02	200
Tortellini with meat sauce	<0.03	<0.03	<3.0	0.2	0.02	200
Ravioli with meat sauce	0.04	<0.03	<3.0	0.2	0.02	200

Table 2.11 Different packaged foodstuffs in metallic containers and metal contamination

Different canned foods and metal contamination

	Average values (mg/kg)				Maximum allowed concentration (mg/kg)			
	Lead	Cadmium	Mercury	Tin	Lead	Cadmium	Mercury	Tin
Fruit products in plain cans								
Fruits salad	0.06	<0.03	–	39	0.10	0.05	(1)	200
Milk powdered in aluminium tubes								
Milk powder	0.01	<0.03	–	–	0.02	(1)	(1)	50
Alcoholic beverages in glass bottles								
Brandy	<0.01	<0.03	–	–	0.20	(1)	(1)	100

(1) There are not official maximum allowed concentration limits with concern to this metal for these food products

Table 2.12 Vegetable products and mercury contamination

Mercury contamination in canned vegetable products

Commercialised food products	Average values ± standard deviation (mg/kg)
Diced tomato in metallic cans	<0.1
Organic diced foods in metallic cans	<0.1
Tomato double paste in aluminium bags	<0.1
Tomato triple paste in plastic bags	<0.1
Dry mushrooms	2.54 ± 1.42
Grinded dry mushrooms	4.69 ± 1.94

- Food flavourings
- Ready products for mashed potatoes, croquettes and potato dumplings
- Minced and dried celery and carrots
- Red *pesto* sauce in glass jars
- Green *pesto* sauce in glass jars
- Cream of black olives in glass jars

Table 2.13 Fish products and mercury contamination

Mercury contamination in canned fish products		
Commercialised fish products	Average values ± standard deviation (mg/kg)	Maximum allowed concentration (mg/kg)
Tuna olive oil in metallic cans	0.28 ± 0.34	1.0
Tuna seed oil in metallic cans	0.46 ± 0.10	1.0
Tuna olive oil in glass jars	0.43 ± 0.36	1.0
Mackerel fillets olive oil in can	<0.1	0.5
Anchovies in glass jars	0.22	0.5
Clams in brine packed in glass jars	<0.1	0.5
Sardines in olive oil packed in glass jars	<0.1	0.5
Smoked salmon packed in plastic films	<0.1	0.5
Fresh cuttlefish packed in plastic films	<0.1	0.5
Fresh cod packed in plastic films	<0.1	0.5
Pasta with anchovies, packed in aluminium tubes	<0.1	0.5

Table 2.14 Semi-processed food products and metal contamination

Heavy metals (mg/kg) in semi-processed foods						
Semi-processed food products on the market	Arsenic	Cadmium	Lead	Cuprum	Zinc	Mercury
Smoked salmon	0.27	<0.01	0.11	<0.5	1.9	<0.10
Tuna in brine	0.10	<0.01	0.02	0.5	7.6	1.01
Cuttlefish	0.51	2.53	0.75	10.1	34	<0.10
Cod	0.21	<0.01	0.19	0.6	5.3	<0.10
Speck	<0.01	<0.01	0.02	1.0	30	<0.10
Sausage	0.02	<0.01	0.01	0.6	8.3	<0.10
Dry mushroom	0.19	0.62	0.61	18	24	1.14
Oregano	(1)	0.45	1.08	(1)	(1)	<0.10
Potato puree	0.01	0.04	0.04	1.5	3.2	<0.10
Peach in pieces	0.01	0.04	0.13	1.2	2.0	<0.10
Blueberries	0.02	0.02	0.05	<0.5	4.0	<0.10
Vegetables	0.02	0.05	0.16	<0.5	4.6	<0.10
Mozzarella cheese	<0.01	<0.01	0.06	<0.5	28	<0.10

(1) This metal has not been detected in analysed samples

- Dried and ground spinach in glass jars
- Ready sauce in glass jars
- Hot sauce in glass jars
- Tomato sauce with basil in glass jars
- Four-cheese Creamy in glass jars
- Minced Courgettes in glass jars

- Mashed potato in bags
- Apples-dried in bags
- Dried peaches in bags
- Dried cranberries in bags
- Diced and sliced organic carrots in glass jars
- Organic garlic paste in glass jars.

Generally, SSICA researchers have found lower values than the maximum allowed limits with the exception of two cans of tuna oil and with regard to Hg content. In particular, dried mushrooms are known to be a 'critical' product.

2.7 Conclusions

Metals have many different applications in food packaging (structural materials, additives, etc.). Current scientific papers report that metal amounts, especially heavy metals, usually comply with legal limits for preserved foodstuffs. Collected data seem to highlight that the influence of FPM is quite limited on condition that positive lists of composition and good manufacturing practices are fully implemented. For example, high concentrations may be due to anomalous corrosion process—this phenomenon still represents the exception—or contaminated raw materials, including also the well-known bioaccumulation. Even recent data on the content of heavy metals in food appear to confirm this conclusion.

References

1. European Commission (2011) Commission Regulation (EU) No. 420/2011 of 29 April 2011 amending Regulation (EC) No. 1881/2006 setting maximum levels for certain contaminants in foodstuffs. Off J Eur Union L111:3–6
2. European Commission (2014) Commission Regulation (EU) No. 488/2014 of 12 May 2014 amending Regulation (EC) No. 1881/2006 as regards maximum levels of cadmium in foodstuffs. Off J Eur Union L138:75–79
3. European Commission (2007) Commission Regulation (EC) No. 333/2007 of 28 March 2007 laying down the methods of sampling and analysis for the official control of the levels of lead, cadmium, mercury, inorganic tin, 3-MCPD and benzo(a)pyrene in foodstuffs. Off J Eur Union L88:29–38
4. European Commission (2011) Commission Regulation (EU) No. 836/2011 of 19 August 2011 amending Regulation (EC) No. 333/2007 laying down the methods of sampling and analysis for the official control of the levels of lead, cadmium, mercury, inorganic tin, 3-MCPD and benzo(a)pyrene in foodstuffs. Off J Eur Union L215:9–16
5. European Parliament and the Council (2004) Regulation (EC) No. 1935/2004 of the European Parliament and of the Council of 27 October 2004 on materials and articles intended to come into contact with food and repealing Directives 80/590/EEC and 89/109/EEC. Off J Eur Union L338:4–17
6. Health Ministry (1973) Decree 21.3.73. Disciplina igienica degli imballaggi, recipienti, utensili, destinati a venire in contatto con le sostanze alimentari o con sostanze d'uso personale. Gazz Uff Repubbl Ital (Supplemento Ordinario) No. 104 of 20.4.73

7. Health Ministry (2007) Decree 18.4.07 n. 76. Regolamento recante la disciplina igienica dei materiali e degli oggetti di alluminio e di leghe di alluminio destinati a venire a contatto con gli alimenti. Gazz Uff Repubbl Ital No. 141 of 20.6.07

8. Health Ministry (2010) Decree 21.12.10 n. 258. Regolamento recante aggiornamento del decreto ministeriale 21 marzo 1973, concernente la disciplina igienica degli imballaggi, recipienti, utensili destinati a venire a contatto con le sostanze alimentari o con sostanze d'uso ersonale, limitatamente agli acciai inossidabili. Gazz Uff Repubbl Ital No. 28 of 4.2.2011

9. Council of Europe (2013) Resolution CM/Res(2013)9 on metals and alloys used in food contact materials and articles. Council of Europe, Strasbourg. Available https://wcd.coe.int/View Doc.jsp?id=2075683&Site=CM. Accessed 15 Oct 2014

10. Cheng J, Ma X, Wu Y (2014) Silica gel chemically modified with ionic liquid as novel sorbent for solid-phase extraction and preconcentration of lead from beer and tea drink samples followed by flame atomic absorption spectrometric determination. Food Anal Methods 7(5):1083–1089. doi:10.1007/s12161-013-9716-3

11. Sorbo A, Turco AC, Di Gregorio M, Ciaralli L (2014) Development and validation of an analytical method for the determination of arsenic, cadmium and lead content in powdered infant formula by means of quadrupole inductively coupled plasma mass spectrometry. Food Control 44:159–165. doi:10.1016/j.foodcont.2014.03.049

12. Talio MC, Alesso M, Acosta MG, Acosta M, Fernández LP (2014) Sequential determination of lead and cobalt in tap water and foods samples by fluorescence. Talanta 127:244–249. doi:10.1016/j.talanta.2014.04.020

13. Oldring PKT, O'Mahony C, Dixon J, Vints M, Mehegan J, Dequatre C, Castle L (2014) Development of a new modelling tool (FACET) to assess exposure to chemical migrants from food packaging. Food Add Contam Part A 31(3):444–465. doi:10.1080/19440049.2013.862348

14. Joint FAO/WHO Expert Committee on Food Additives (2006) Evaluation of certain food additives and contaminants. In: 67th meeting, Rome. WHO technical report series, vol 940, pp 1–94, 20–29 June 2006. Available http://whqlibdoc.who.int/trs/WHO_TRS_940_eng.pdf. Accessed 15 Oct 2014

15. EFSA AFC (2008) Scientific opinion of the panel on food additives, flavourings, processing aids and food contact materials on a request from european commission on safety of aluminium from dietary intake. EFSA J 754:1–34. http://www.efsa.europa.eu/en/efsajournal/doc/754.pdf. Accessed 24 Sep 2014

16. European Commission (2012) Commission Regulation (EU) No. 380/2012 of 3 May 2012 amending Annex II to Regulation (EC) No. 1333/2008 of the European Parliament and of the Council as regards the conditions of use and the use levels for aluminium-containing food additives. Off J Eur Union L119:14–38

17. CDC (2012) CDC response to advisory committee on childhood lead poisoning prevention. In: Recommendations in 'low level lead exposure harms children: a renewed call of primary prevention', June 2012. Centers for Disease Control and Prevention, Atlanta. Available http://www.cdc.gov/nceh/lead/acclpp/cdc_response_lead_exposure_recs.pdf. Accessed 15 Oct 2014

18. Codex Alimentarius Commission (1985) General standard for contaminants and toxins in food and feed, CODEX STAN 193-1995, last amendment: 2013. In: World Health Organization/Food and Agriculture Organization of the United Nations, Rome. Available http://www.codexalimentarius.org/download/standards/17/CXS_193e.pdf. Accessed 15 Oct 2014

19. Lopez E, Figueroa S, Oset-Gasque MJ, Gonzalez MP (2003) Apoptosis and necrosis: two distinct events induced by cadmium in cortical neurons in culture. Br J Pharmacol 138(5):901–911. doi:10.1038/sj.bjp.0705111

20. Contam EFSA (2009) Scientific opinion on arsenic in food. EFSA J 7(10):1351–1550. doi:10.2903/j.efsa.2009.1351

21. European Commission (2004) Commission Regulation (EC) No. 242/2004 amending Regulation (EC) No. 466/2001 as regards inorganic tin in foods. Off J Eur Union L42:3–4

22. Technical Committee ECISS/TC 26 (2005) EN 102333:2005. Steel for packaging. Flat steel products intended for use in contact with foodstuffs, products or beverages for human and animal consumption. Tin coated steel (tinplate). European Committee for Standardization, Brussels

23. Health Ministry (1984) Decree 18.2.84. Disciplina dei contenitori in banda stagnata saldati con lega stagno-piombo ed altri mezzi. Gazz Uff Repubbl Ital No. 76 of 16.3.84

24. Health Ministry (1988) Decree 1.6.88 n. 243. Disciplina degli oggetti in banda cromata verniciata destinati a venire in contatto con gli alimenti. Gazz Uff Repubbl Ital No. 153 of 01.7.88

25. Technical Committee CEN/TC 132 (2004) EN 601:2004. Aluminum and aluminium alloys. Castings. Chemical composition of castings for use in contact with foodstuff. European Committee for Standardisation, Brussels

26. Technical Committee CEN/TC 132 (2004) EN 602:2004. Aluminium and aluminium alloys. Wrought products. Chemical composition of semi-finished products used for the fabrication of articles for use in contact with foodstuff European Committee for Standardisation, Brussels

27. Technical Committee CEN/TC 132 (2004) EN 14287:2004. Aluminium and aluminium alloys—Specific requirements on the chemical composition of products intended to be used for the manufacture of packaging and packaging components. European Committee for Standardisation, Brussels

28. Technical Committee CEN/TC 261 (2000) EN 13046:2000. Packaging. Flexible cylindrical metallic tubes. Dimensions and tolerances. European Committee for Standardisation, Brussels

29. European Commission (2007) Commission Directive 2007/42/CE of 29 June 2007 relating to materials and articles made of regenerated cellulose film intended to come into contact with foodstuffs. Off J Eur Union L172:71–82

30. European Commission (2011) Commission Regulation (EU) No. 10/2011 of 14 January 2011 on plastic materials and articles intended to come into contact with food. Off J Eur Union L12:1–89

31. European Commission (2009) Commission Regulation (EC) No. 450/2009 of 29 May 2009 on active and intelligent materials and articles intended to come into contact with food. Off J Eur Union L135:3–11

32. Samal SK, Fernandes EG, Corti A, Chiellini E (2014) Bio-based polyethylene-lignin composites containing a pro-oxidant/pro-degradant additive: preparation and characterization. J Polym Environ 22(1):58–68. doi:10.1007/s10924-013-0620-0

33. Roman MJ, Tian F, Decker EA, Goddard JM (2014) Iron chelating polypropylene films: manipulating photoinitiated graft polymerization to tailor chelating activity. J Appl Polym Sci 131(4). doi:10.1002/app.39948

34. EFSA Scientific Committee (2011) Guidance on the risk assessment of the application of nanoscience and nanotechnologies in the food and feed chain. EFSA J 9(5):2140–2176. doi:10.2903/j.efsa.2011.2140

35. EFSA AFC (2012) Scientific opinion on the safety evaluation of the substance, titanium nitride, nanoparticles, for use in food contact materials. EFSA J 10(3):2641–2649. doi:10.2903/j.efsa.2012.2641

36. Martirosyan A, Schneider YJ (2014) Engineered nanomaterials in food: implications for food safety and consumer health. Int J Environ Res Public Health 11(6):5720–5750. doi:10.3390/ijerph110605720

37. Lin QB, Li H, Zhong HN, Zhao Q, Xiao DH, Wang ZW (2014) Migration of Ti from nano-TiO_2-polyethylene composite packaging into food simulants. Food Addit Contam Part A 31(7):1284–1290. doi:10.1080/19440049.2014.907505

38. Ysart G, Miller P, Croasdale M, Crews H, Robb P, Baxter M, De L'Argy C, Harrison N (2000) 1997 UK total diet study dietary exposures to aluminium, arsenic, cadmium, chromium, copper, lead, mercury, nickel, selenium, tin and zinc. Food Addit Contam 17(9):775–786. doi:10.1080/026520300415327

39. Hadiani MR, Farhangi R, Soleimani H, Rastegar H, Cheraghali AM (2014) Evaluation of heavy metals contamination in Iranian foodstuffs: canned tomato paste and tomato sauce (ketchup). Food Addit Contam Part B Surveill 7(1):74–78. doi:10.1080/19393210.2013.848384

40. Chahid A, Hilali M, Benlhachimi A, Bouzid T (2014) Contents of cadmium, mercury and lead in fish from the Atlantic sea (Morocco) determined by atomic absorption spectrometry. Food Chem 147:357–360. doi:10.1016/j.foodchem.2013.10.008

41. Cherfi A, Abdoun S, Gaci O (2014) Food survey: levels and potential health risks of chromium, lead, zinc and copper content in fruits and vegetables consumed in Algeria. Food Chem Toxicol 70:48–53. doi:10.1016/j.fct.2014.04.044

42. Illuminati S, Annibaldi A, Truzzi C, Scarponi G (2014) Recent temporal variations of trace metal content in an Italian white wine. Food Chem 159:493–497. doi:10.1016/j.foodchem.2014.03.058

Chapter 3
Plasticisers Used in PVC for Foods: Assessment of Specific Migration

Luciana Bolzoni

Abstract The use of polyvinyl chloride (PVC) in food packaging is mainly related to the plasticity of the same material when used in wrapping films and in gaskets for metal closures (applications: glass jars and bottles). Anyway, main required characteristics are the flexibility, the softness and the possibility of being used for wrapping films and hermetic closures. Pure PVC is a rigid material, but it may also be mixed in remarkable proportions with other substances: the final product may become flexible, soft and plastic. Many plasticisers may be used in the European Union in accordance with the Regulation (EU) No. 10/2011 on plastic materials and articles intended to come into contact with food. In relation to food contact-approved PVC materials, inglobated plasticisers can gradually migrate from the plasticised object to foods depending on the influence of factors such as the temperature or the physical medium (solvent, food). The Regulation (EU) No. 10/2011 provides specific migration limits for different plasticisers. The analytical control of these limits in foods and/or in food simulants is important by the viewpoint of food safety. Currently available and used methods for the evaluation of specific migration are reviewed in this paper.

Keywords Gas chromatography · High-performance liquid chromatography · Mass spectrometry · Phthalate · Plasticiser · Polyvinyl chloride · QuEChERS · Specific migration limit

Abbreviations

acPG	Acetylated partial glycerides
AFC	Panel on food additives, flavouring, processing aids and materials in contact with food
ATBC	Acetyl tributyl citrate
BBP	Benzyl butyl phthalate
BBPd4	Benzyl butyl phthalate deuterated
DEHA	Bis-ethylhexyl adipate
DINCH	1,2-Cyclohexane dicarboxylic acid, diisononyl ester

© The Author(s) 2015

C. Barone et al., *Food Packaging Hygiene*, Chemistry of Foods,
DOI 10.1007/978-3-319-14827-4_3

43

DBP	Dibutyl phthalate
DBPd4	Dibutyl phthalate deuterated
DBS	Dibutyl sebacate
DCP	Dicyclohexyl phthalate
DEHP	Di(2-ethylhexyl) phthalate
DEHPd4	Di(2-ethylhexyl) phthalate deuterated
DEP	Diethyl phthalate
DEPd4	Diethyl phthalate deuterated
DEHS	Di-2-ethylhexyl sebacate
DIBP	Diisobutyl phthalate
DIDP	Diisodecyl phthalate
DINP	Diisononyl phthalate
DMG	Dimethyl glutarate
DMP	Dimethyl phthalate
DMPi	Dimethyl pimelate
DMS	Dimethyl sebacate
DPP	Di-n-propyl phthalate
DOP	Dioctyl phthalate
DOPd4	Dioctiyl phthalate deuterated
DpeP	Dipentyl phthalate
ESBO	Epoxidised soybean oil
EFSA	European Food Safety Authority
SSICA	Experimental Station for the Food Preserving Industry
FID	Flame ionisation detector
GC	Gas chromatography
GC/MS	Gas chromatography/mass spectrometry
HPLC	High-performance liquid chromatography
LC	Liquid chromatography
LC/MS	Liquid chromatography/mass spectrometry
LC/MS/MS	Liquid chromatography/tandem mass spectrometry
MW	Molecular weight
MgSO$_4$	Magnesium sulphate
PVC	Polyvinyl chloride
PSA	Primary and secondary amines
QuEChERS	Quick, Easy, Cheap, Effective, Rugged and Safe
RPLC	Reversed-phase liquid chromatography
SML	Specific migration limit
THF	Tetrahydrofuran

3.1 Introduction

Polyvinyl chloride (PVC) is one of the most used materials in the world; the use of this polymer in the production of objects dates back to the first half of the twentieth century. The use of PVC in contact with foods—for example in pipes,

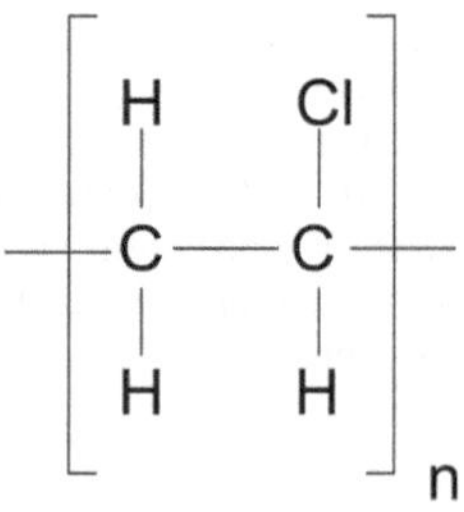

Fig. 3.1 Chemical structure
of polyvinyl chloride (PVC).
BKchem version 0.13.0, 2009
(http://bkchem.zirael.org/
index.html) has been used for
drawing this structure

conveyor belts of food industry or in packaging materials—is mainly due to the intrinsic plasticity of the polymer. This feature, also named 'flexibility', and softness are fundamental features for the seals of caps, for the hermetic closure of glass jars and bottles when destined to contain foods. In particular, hermeticity is essential because of the need of assuring the sanitary safety of processed foods after thermal treatments such as pasteurisation or sterilisation.

Pure PVC (Fig. 3.1) is a rigid material, and it is used in several ways. In detail, PVC may be used alone in extrusion processes for the production of section bars (doors and windows, pipes, etc.). On the other side, PVC can be also mixed with plasticizing substances in high proportions. The final product may be softer, more flexible and with remarkable plastic properties if compared with a pure PVC object.

In relation to most used plasticisers for PVC, a peculiar group concerns polycarboxylic acids: phthalic acid esters, adipic acid esters, sebacic acid esters and so on, with alcohols of variable length.

By the chemical viewpoint, plasticisers can be inglobated in PVC because of the solvation of C–CI polar chemical bonds on the polymeric chain. The solvation is mainly due to carboxylic (COO^-) polar groups; should aromatic rings be attached to the chemical structure, the influence of π-electrons would be an additional factor. A physical bond is hence created between PVC and plasticisers; however, this bond cannot be confused with a chemical interaction. For this reason, the plasticizing agent can gradually migrate from food contact PVC surfaces into packaged foods in connection with concurring factors such as storage temperatures and/or the chemical and physical nature of the medium (solvent, food).

3.1.1 Phthalates

A group of plasticisers that has been repeatedly discussed in the last few years concerns phthalates. Essentially, they are phthalic acid esters (Fig. 3.2): some of these molecules are more used than others. As a consequence, the abundance of the scientific literature concerns only a limited subgroup of phthalates.

One of the most used phthalates (as PVC plasticizer) is di-2-ethylhexyl phthalate (DEHP). This molecule is a phthalic acid ester obtained with 2-ethylhexanol.

With reference to the human health, main doubts on this type of phthalates concern the role of DEHP as destroyer of the human endocrine system; in particular, detrimental effects of DEHP have been detected on the reproductive system [1].

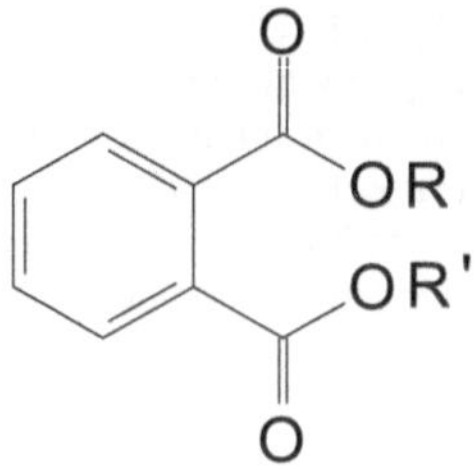

Fig. 3.2 General formula of
phthalates. R e R' group can
be equal. BKchem version
0.13.0, 2009 (http://bkchem.
zirael.org/index.html) has
been used for drawing this
structure

Several restrictions on DEHP and some phthalates have been introduced in the
European Union since several years. In fact, the use of phthalates is not allowed
with concentrations higher than 0.1 %, neither in toys nor in childhood products;
the reason of this restriction is due to exposure hazards that can originate from
chewing or sucking such objects for a long time [2].

Some restrictions also include materials that are in contact with food. The
Regulation (EU) No. 10/2011 [3] defines restrictions for the use and specific
migration limits (SML) for some types of phthalates on the basis of the tolerable
daily intake (TDI) [1, 4–7]:

- Benzyl butyl phthalate (BBP)
- Dibutyl phthalate (DBP)
- Diisodecyl phthalate (DIDP)
- Diisononyl phthalate (DINP)
- DEHP.

Other phthalates are not taken into account such as diisobutyl phthalate (DIBP):
this substance is widely used and, therefore, easily detectable in analytical samples
and blanks.

Consequently, SML values lower than 0.010 mg/kg are accepted, according to
the Regulation No. 10/2011, for substances which are not included in Annex I.

3.1.2 Epoxidised Soybean Oil

Epoxidised soybean oil (ESBO) (Fig. 3.3) is widely used as an alternative to
phthalates because of similar features in relation to plasticizing effects. Moreover,
ESBO can also act as a scavenger for hydrochloric acid, which is released from
PVC during heat treatments of PVC-based materials. For this reason, ESBO is
definable as a 'decontaminating' agent when used in connection in PVC.

However, ESBO is highly lipophilic: the notable attitude to migration (or
release from organic supports) of this natural origin plasticiser has been outlined
about 10 years ago. In relation to available studies, two researches—an European
study on baby foods [8] and other tests on oily foods in jars—have to be mentioned [9]. In detail, the release of ESBO in oils after the free contact between

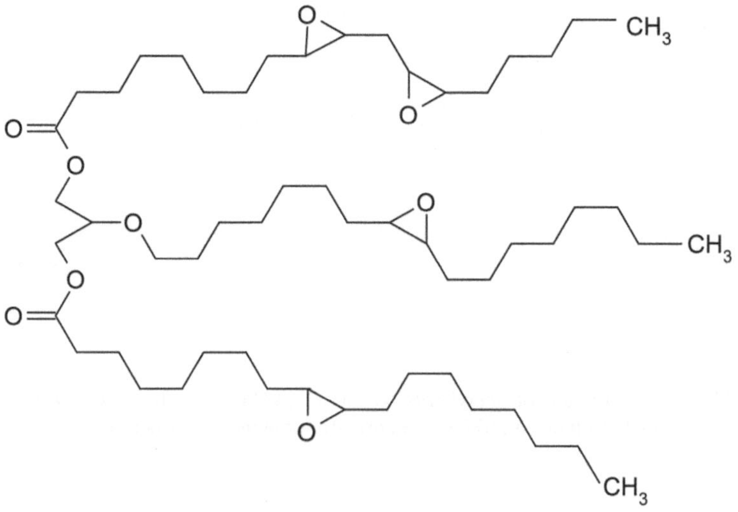

Fig. 3.3 Chemical structure of epoxidised soybean oil (*ESBO*). BKchem version 0.13.0, 2009 (http://bkchem.zirael.org/index.html) has been used for drawing this structure

foods and plasticized materials (ESBO was the only plasticizing agents) led SML and global limits to notable levels.

After the evaluation of the European Food Safety Authority (EFSA), the European Community has established a SML value of 30 mg/kg for baby food. On the other side, SML has been raised to 300 mg/kg for other foods in relation to ESBO for a limited period. The main reason was correlated to the temporal implementation: in other words, manufacturers of films and sealing compounds should have been reasonably able to find other plasticisers and formulations within a specified deadline. Subsequently, a SML value of 60 mg/kg, which is still effective, has been established [3].

3.1.3 Other Monomeric Plasticisers

Acetyl tributyl citrate (ATBC) (Fig. 3.4) is a common plasticiser for food contact applications. This molecule shows an excellent biodegradability and a chemically remarkable affinity to PVC because of the polarity. The Regulation (EU) No. 10/2011 does not report SML values for ATBC; consequently, a general SML of 60 mg/kg is considered. However, ATBC is also highly soluble in lipophilic materials and, therefore, highly soluble in them. Sebacates are also considered good plasticisers, and their moderate toxicity is known at present; on the other hand, dibutyl sebacate (DBS) (Fig. 3.5) only is currently reported in the 'positive list' of the Regulation (EU) No. 10/2011 without limits of specific migration for food contact applications (SML = 60 mg/kg).

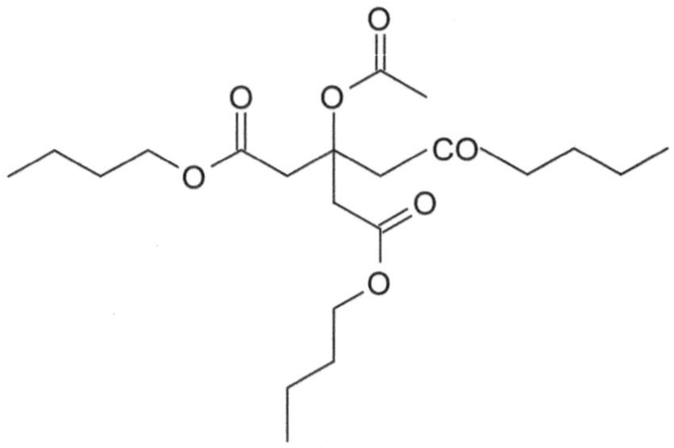

Fig. 3.4 Chemical structure of acetyl tributyl citrate (*ATBC*). BKchem version 0.13.0, 2009 (http://bkchem.zirael.org/index.html) has been used for drawing this structure

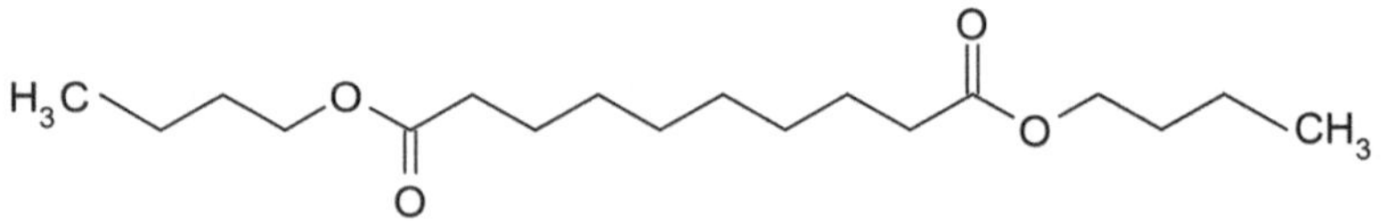

Fig. 3.5 Chemical structure of dibutyl sebacate (*DBS*). BKchem version 0.13.0, 2009 (http://bkchem.zirael.org/index.html) has been used for drawing this structure

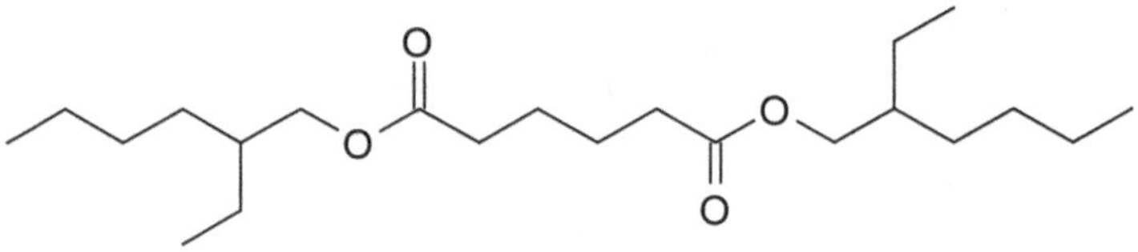

Fig. 3.6 Chemical structure of bis-ethylhexyl adipate (*DEHA*). BKchem version 0.13.0, 2009 (http://bkchem.zirael.org/index.html) has been used for drawing this structure

Another widely used plasticiser is bis-ethylhexyl adipate (DEHA) (Fig. 3.6); the above-mentioned Reg. No 10/2011 defines a SML values for DEHA of 18 mg/kg on the basis of the opinion of the Scientific Committee on Food in 2000 [10].

1,2-Cyclohexane dicarboxylic acid, diisononyl ester (DINCH) (Fig. 3.7) is currently considered the non-aromatic alternative to phthalates [11] and to DEHP in particular. However, DINCH is reported to show a high migration tendency when in contact with oily foods. Some acetylated partial glycerides (acPG)—monoglycerides and diglycerides—are employed as plasticisers, for example, glyceril monolaurate diacetate, glyceril dilaurate monoacetate hydrogenated castor oil with glycerine and acetic acid (ARMG). In relation to acPG, the Regulation (EU) No. 10/2011 does not impose a SML value; as a result, the limit of 60 mg/kg is considered. acPG are considered as plasticisers without particular safety consequences because of their

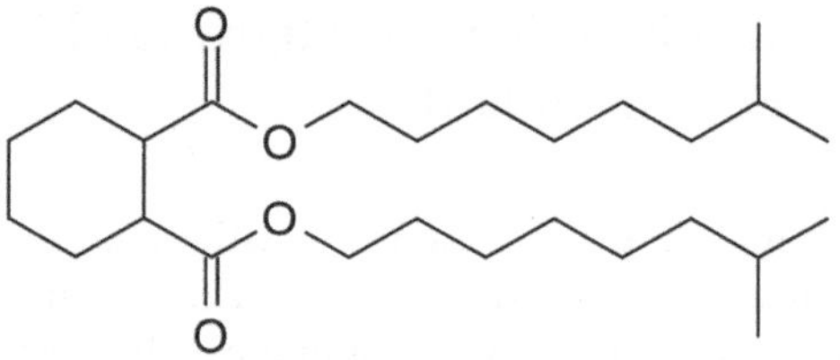

Fig. 3.7 Chemical structure of 1,2-cyclohexane dicarboxylic acid diisononyl ester (*DINCH*). BKchem version 0.13.0, 2009 (http://bkchem.zirael.org/index.html) has been used for drawing this structure

chemical similarity to natural raw materials. The presence of long alkyl chains is the main cause of the notable solubility of acPG in non-polar solvents; therefore, a potentially high migration in oil can be hypothesised.

3.1.4 Polyadipates

Monomeric plasticisers have been replaced at least partly by polymeric plasticisers. In particular, polyadipates are considered as good options: their high molecular weight does not help migration. However, molecules up to 1,000 Da migrate more easily, and a migration limit of 30 mg/kg has been defined for these situations, on condition that polyadipates are authorised by the Reg. No 10/2011. Figure 3.8 shows some structural elements of polyadipates [3].

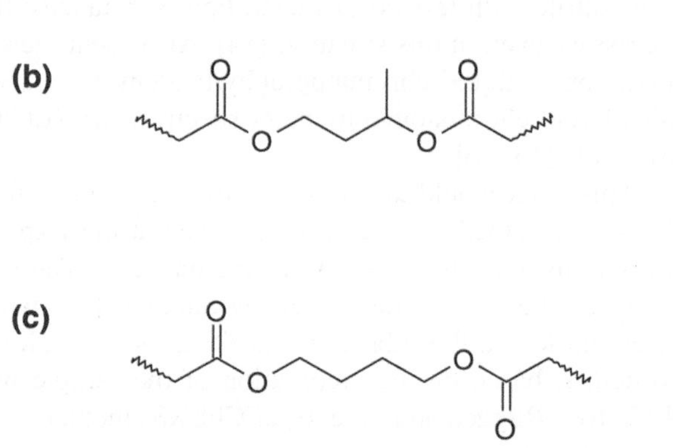

Fig. 3.8 Structural elements of polyadipates: 1,2-propanediol adipate (**a**), 1,3-butanediol adipate (**b**) and 1,4-butanediol adipate (**c**). BKchem version 0.13.0, 2009 (http://bkchem. zirael.org/index.html) has been used for drawing these structures

3.2 Analytical Controls of Specific Migration Limits in Foods and Food Simulants

As already outlined for all the mentioned plasticisers, some SML values are required by the Regulation (EU) No. 10/2011. These numerical restrictions are referred to food products for the whole commercial life. SML must be checked and estimated also in simulant liquids on condition that the material has not been yet in contact with the food, and contact conditions are established in the Chap. 2 of Annex V of the Regulation [3].

Consequently, there is the necessity of available laboratories committed to the control and the verification of food safety in relation to packaged foods and packaging materials containing plasticisers. In detail, the creation and the implementation of reliable analytical methods for the analysis of plasticisers in packaging materials is needed. On the basis of chemical features of above-mentioned substances, the most important food matrices are obviously oily foods including the 'D simulant' (oil) [3]. By the analytical viewpoint, the main reason is correlated to observed releases: the migration is higher towards oil, greasy and oily foods.

3.2.1 Methods for the Analysis of Phthalates

Published methods are based, for the most part, on the instrumental determination in gas chromatography/mass spectrometry (GC/MS) after extraction or purification of the extract.

The extraction is often carried out with low polar solvents, and plasticisers are extracted together with lipid substances from which they have to be separated before gas chromatography (GC) determination. To this aim, gel permeation chromatography has been used [12]. Direct injection without purification has successfully been used, using a special GC injector with inner thermal desorption and external discharge of lipid substances [13].

The extraction can also be carried out with polar solvents such as methanol or acetonitrile with less co-extracted lipid substances; however, a purification step is necessary even in this situation [14]. At present, new methods based on the determination in liquid chromatography/tandem mass spectrometry (LC/MS/MS) and also liquid chromatography/mass spectrometry (LC/MS) with high resolution are available [15, 16].

This work would also describe two procedures for the determination of phthalates. These methods are currently used at the Experimental Station for the Food Preserving Industry (SSICA) in Parma, Italy. The first procedure is based essentially on the method developed by Sannino [17] which requires the extraction in acetonitrile and the subsequent purification on a Florisil column, while the second system is based on the preparation of the sample with the 'Quick, Easy, Cheap, Effective, Rugged and Safe' (QuEChERS) method.

The QuEChERS method is an easy and effective technique for the multi-residual analysis of pesticides in food, allowing also the considerable reduction of time and costs [18]. This procedure has been tested at SSICA for the analysis of phthalates in oil and products in oil.

3.2.2 QuEChERS Method: A Case Study

The preparation of food samples with the QuEChERS method has been schematically reported in Table 3.1. This section concerns the description of a normal determination of phthalates in oils and anchovies in oils at SSICA.

In relation to this research, the instrumental determination has been based on a triple quadrupole GC/MS system in series. The used column has been a Phenomenex Guardian^{TM} column (30 m × 0.25 mm inner diameter; film thickness 0.25 μm) with following conditions: injection temperature: 250 °C; source temperature: 230 °C; transfer line temperature: 300 °C. Table 3.2 shows analysed phthalates and deuterated phthalates used as internal standards. In addition, chromatograms of a standard mixture of analysed phthalates and of used internal deuterated standards phthalates have been reported in Figs. 3.9 and 3.10.

Ten recovery trials have been tested in oil and anchovies in oil at the 0.5 mg/kg fortification level for all phthalates except for DEHP and dioctyl phthalate (DOP) with 1 mg/kg and DIDP with 10 mg/kg.

Mean per cent recoveries have been higher than 77 % in oil with the exception of dicyclohexyl phthalate (DCP) and DIDP with 60 %. With concern to anchovies in oil, all recoveries have been higher than 73 %.

Table 3.1 A QuEChERS method scheme

Step number	Description
Step 1	Weight 5 g of homogenised sample
Step 2	Add internal standards and 10 ml of acetonitrile. Shake for 1 min
Step 3	Add the extraction salt: 4 g of magnesium sulphate (MgSO$_4$) + 1 g of sodium chloride + 1 g of trisodium citrate dehydrate + 0.5 g of disodium hydrogen citrate. Shake for 1 min
Step 4	Centrifuge for 5 min at 4,000 rpm
Step 5	Transfer 6 ml in the test tube for the purification of vegetal matrices containing 25 mg primary and secondary amines (PSA) and 150 mg of MgSO$_4$. Should fat matrices be analysed, transfer 6 ml in the test tube for the purification of matrices containing 150 mg of PSA, 150 mg of C18 sorbent and 900 mg of MgSO$_4$
Step 6	Shake with vortex for 1 min
Step 7	Centrifuge for 5 min at 4,000 rpm Transfer in vial for the analysis in GC/MS

This procedure is used at SSICA with concern to the preparation of samples for the subsequent analysis of phthalates

Table 3.2 Phthalates and deuterated phthalates

Phthalates	Precursor ions (m/z)	Product ions (m/z)
Dimethyl phthalate (DMP)	163	133 + 135 + 105
Diethyl phthalate (DEP)	149	65 + 93 + 121
Diethyl phthalate deuterated (DEPd4)	153	69 + 97 + 125
Di-*n*-propyl phthalate (DPP)	149	65 + 93 + 121
Diisobutyl phthalate (DIBP)	149	65 + 93 + 121
Dibutyl phthalate (DBP)	149	65 + 93 + 121
Dipentyl phthalate (DpeP)	149	65 + 93 + 121
Dibutyl phthalate deuterated (DBPd4)	153	69 + 97 + 125
Benzyl butyl phthalate (BBP)	149	65 + 93 + 121
Dicyclohexyl phthalate (DCP)	149	65 + 93 + 121
Benzyl butyl phthalate deuterated (BBPd4)	153	69 + 97 + 125
Di(2-ethylhexyl)phthalate (DEHP)	149	65 + 93 + 121
Di(2-ethylhexyl)phthalate deuterated (DEHPd4)	153	69 + 97 + 125
Dioctyl phthalate (DOP)	149	65 + 93 + 121
Diisononyl phthalate (DINP)	149	65 + 93 + 121
Diisodecyl phthalate (DIDP)	149	65 + 93 + 121
Dioctyl phthalate deuterated (DOPd4)	153	69 + 97 + 125

Precursor ions and product ions

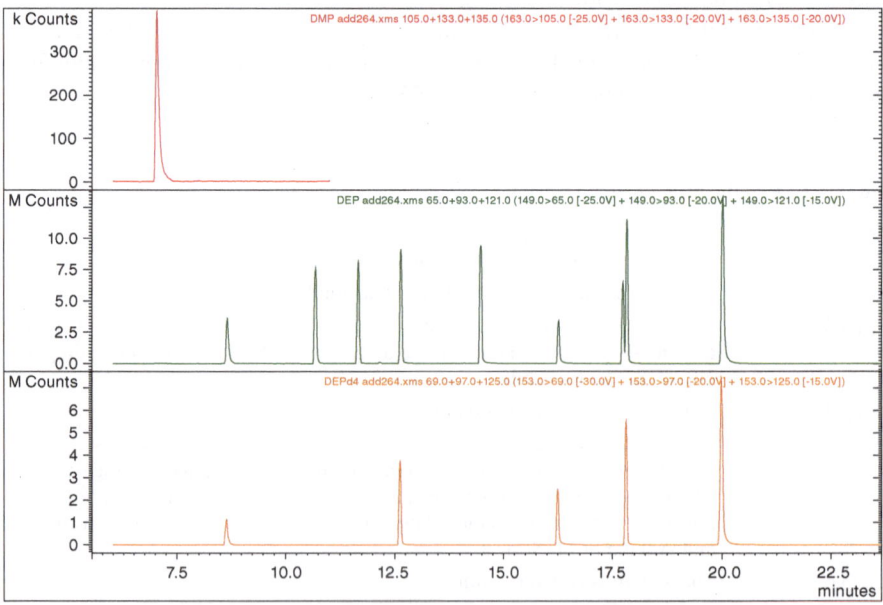

Fig. 3.9 Chromatograms of product ions used for the quantification: mixture of standard phthalates. In order: DMP (*red line*), DEP, DPP,DIBP, DBP, DPeP, BBP, DCP, DEHP, DOP (*green line*), DEPd4, DBPd4, BBPd4, DEHPd4 and DOPd4 (*orange line*)

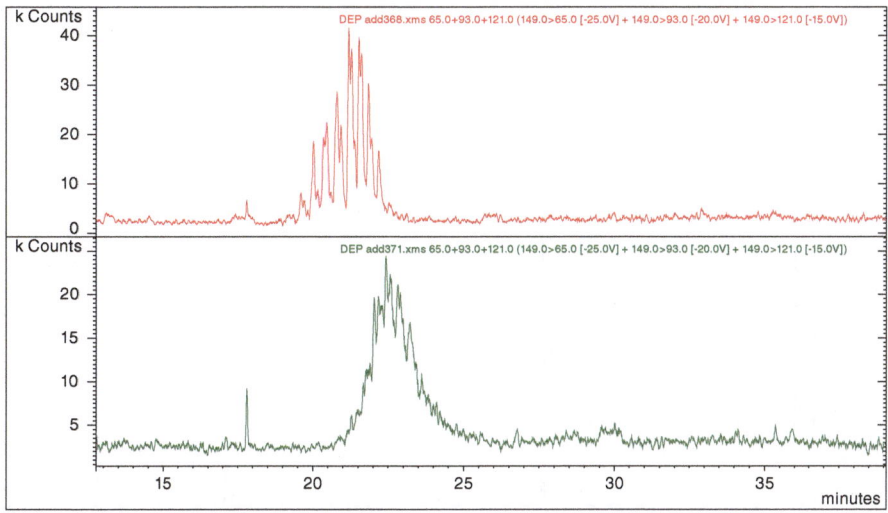

Fig. 3.10 Chromatograms of DINP (*red line*) and DIDP (*green line*) standards

Measured relative standard deviations for all analysed phthalates have been lower than 7.5 and 17 % in oil and anchovies in oil, respectively.

3.2.3 Analysis Methods for ATBC, DBS, DEHA, DINCH, Mono and Partially Acetated Diglycerides of Fatty Acids

Usually, methods used for the analysis of phthalates in foods can be also chosen when speaking of other non-polymeric plasticisers such as ATBC, DBS, di-2-ethylhexyl sebacate (DEHS), DEHA, DINCH and acPG.

Complex mixtures of non-polymeric plasticisers—including phthalates, DEHA, DINCH, DBS, ATBC, DEHS, glycerol monolaurate acetate and mono glyceryl myristate—can be determined in GC/MS with thermal desorption injector [13].

It has been reported also that the method of analysis of phthalates can be used with success when speaking of DBS, ATBC, DEHA and DINCH [17]. This procedure has been later modified in the clean-up step by means of the elution with more polar solvents in order to determine also acPG [17]. In Fig. 3.11 are reported GC/MS chromatograms of eight monomeric plasticisers in oil.

On the other side, recoveries are not suitable for all these monomeric plasticisers when the QuEChERS method is considered. In detail, mean recoveries appear good for DBS, ATBC, DEHA and ARMG with relative standard deviations ≤15 %.

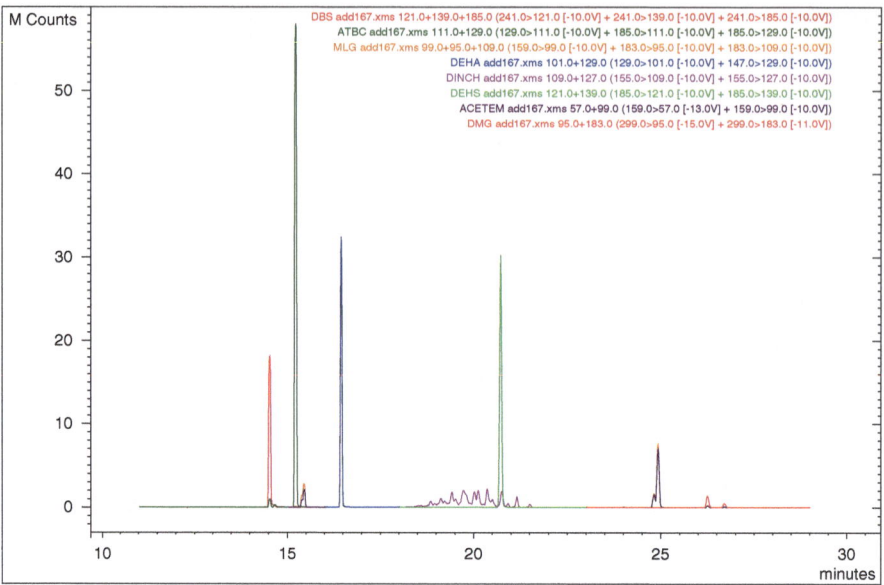

Fig. 3.11 Chromatograms of eight different monomeric plasticisers in oil samples, including DBS (*red line*), ATBC (*dark green line*), DEHA (*blue line*), DINCH (*purple line*), DEHS (*green line*) and DMG (*red line*, on the extreme right)

3.2.4 Analytical Method for ESBO in Foods

Almost all the methods used for the analytical determination of ESBO in foods derive from the method of Castle et al. [19], based on the determination in GC/MS of the methyl ester of diepoxylinoleic acid. This procedure requires the addition of one ester of the diepoxidised fatty acid as the internal standard, followed by the lipidic extraction of food samples and the transmethylation in alkaline conditions [19]. The lipidic extract is suitable for GC/MS analysis without further clean-up. Methylated epoxy fatty acids are then derivatised with cyclopentanone to form 1,3 dioxolanes which are successively determined by capillary GC/MS monitoring the single ions [19].

The Castle method has been used for a study concerning the content of ESBO in child foods [8]. In addition, it has been reported that lipid extraction and dioxolane derivatisation steps may be successfully avoided [20]. In fact, lipids may be transesterified in methyl esters directly into food samples with optimised reaction times, so that the water content does not disturb and avoids saponification [20]. Then, methyl esters of diepoxidised fatty acids are isolated through a normal phase in high-performance liquid chromatography (HPLC) before being transferred online to a GC/flame ionisation detector (FID) system. This method has been also modified by means of the elimination of LC preseparation using a polar column for GC (detector: MS), even if it is still possible to use FID in many cases [21].

A different procedure based on the use of a thermospray interface and reversed-phase liquid chromatography (RPLC) with a serial MS has been reported [22]. This method requires the extraction with dichloromethane of samples without further clean-up. The chromatographic separation is obtained through two C18 columns and mobile phase with the blend of acetic acid, acetone and acetonitrile in gradient conditions [22].

The procedure used at SSICA has been derived from Castle et al. [19], even if it has been noticeably modified in relation to the extraction step, as reported below.

In detail, food samples have to be considered in the following way:

(a) For foods with fat content from 0 to 5 %, the sample should be 30 g
(b) For products with fat content >5 %, the sampled food should allow the extraction of 1–1.5 g of lipids at least; the addition of water is needed in order to obtain 30 g of sample.

Subsequently, the addition of the solution of internal reference should be made in such a quantity that a concentration of 1 mg/g of fat in samples is obtained. Alternatively, 0.5 mg of solution of internal reference should be added when analysing a fat-free sample.

The introduction of a weighted sample in a beaker or in a container suitable for homogenisers is needed; next steps concern the following:

- The addition of 150 ml of hexane–acetone blend 50/50 for 10 min at 9,000 rpm
- The filtration on Buchner funnels with glass fibre filters
- The recovery of filtrates in a separating funnel and the separation between phases (aqueous phase can be eliminated)
- The collection of the organic phase in a flask containing 50 g of anhydrous sodium sulphate.

Subsequently, analysts have to:

(a) Go on with the filtration on a glass fibre filter washing sodium sulfate with portions of solution repeatedly
(b) Evaporate with the rotary evaporator and later under nitrogen, in order to eliminate any trace of solvent and only have the residual fat matter extracted from the sample.

Should the analysis be carried out on a food simulant, the following procedure has to be considered.

(a) Transfer 0.50 ± 0.02 g of D simulant (oil) or fat matter obtained through extraction (or the whole extract for non-fat samples) in a test tube. When speaking of oil simulant, add 0.5 ml of internal standard solution and evaporate the solvent under nitrogen
(b) Add 1 g of anhydrous sodium sulphate and 2 ml of methoxide sodium, shake and put in the stove at 65 °C for two hours shaking the test tube from time to time. After two hours, the extract in the test tube must be clear and homogeneous; should two phases are still observable, transmethylation would necessarily be repeated (the extract is not sufficiently dry)

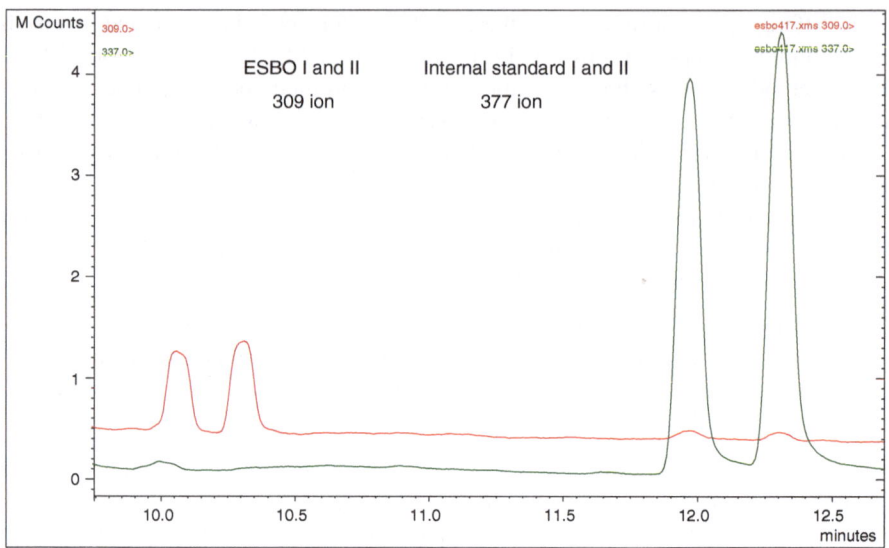

Fig. 3.12 Chromatograms of ESBO target ions (*red line*) and internal standard (*green line*) in olive oil

(c) Put the test tube under the stream of nitrogen and evaporate completely the solvent. Add in order: 2.5 ml of isooctane, 15 ml of cyclopentanone, 250 μl of boron trifluoride and immediately shake for 30 s. Add 5 ml of sodium chloride shaking again for 30 s

(d) Let stand to let separate phases and decant the supernatant liquid that will be possibly stored in freezer before being injected in the GC/MS apparatus, monitoring single ions. The standards for the calibration curve are added to oil simulant in order to have a behaviour similar to the sample; then, methylation and derivatisation are carried out like the sample. Figure 3.12 shows the chromatographic traces of ESBO target ions (309 m/z) at a concentration of 20 mg/kg and of the internal standard in oil (377 m/z).

With the above-described method, repeatability trials have been carried out on olive oil with the addition of 40 mg/kg of ESBO (10 tests) and on meat baby foods with the addition of 20 mg/kg of ESBO. Recoveries have been higher than 95 %; standard deviation values have been 6.7 and 8.9 % for olive oils and baby foods respectively.

3.2.5 Analysis of Polyadipates

The analytical determination of polyadipates is rather complex because of the lack of a single analytical reference substance.

A reliable method concerns [23] the determination of polyadipates through the extraction of foods with acetone/hexane 1:1 v/v, the subsequent transmethylation with etherated boron trifluoride/methanol, the clean-up procedure by size-exclusion chromatography and the final analysis of dimethyl adipate obtained by GC/MS.

The calibration has been performed submitting portions of polyadipate solution used as plasticiser in the food packaging, analysed with the same procedure of transmethylation and submitted to the clean-up used for the sample. This behaviour presumes that migrated polyadipates (small molecules) contain the same proportions of adipate present in the total polyadipate.

The analysis of polyadipates is carried out at the SSICA with the method suggested by Biedermann and Grob [24]. In addition, this procedure is based on the 'adipate' measurement unit. The food or simulating liquid is subject—after dissolution in tetrahydrofuran (THF)—to transesterification with sodium butoxide. Should polyadipate be present, it would be turned into dibutyl adipate: this molecule has a better chromatographic behaviour if compared with dimethyl adipate when obtained with another method [23].

The calibration is performed submitting portions of solution of dimethyl adipate, added of weighed quantities of plasticizer-free oil, to the same procedure of transesterification of the sample. A chromatogram of oil (D simulant) is shown in Fig. 3.13: the simulant has been put in contact with a PVC capsule plasticised with polyadipates.

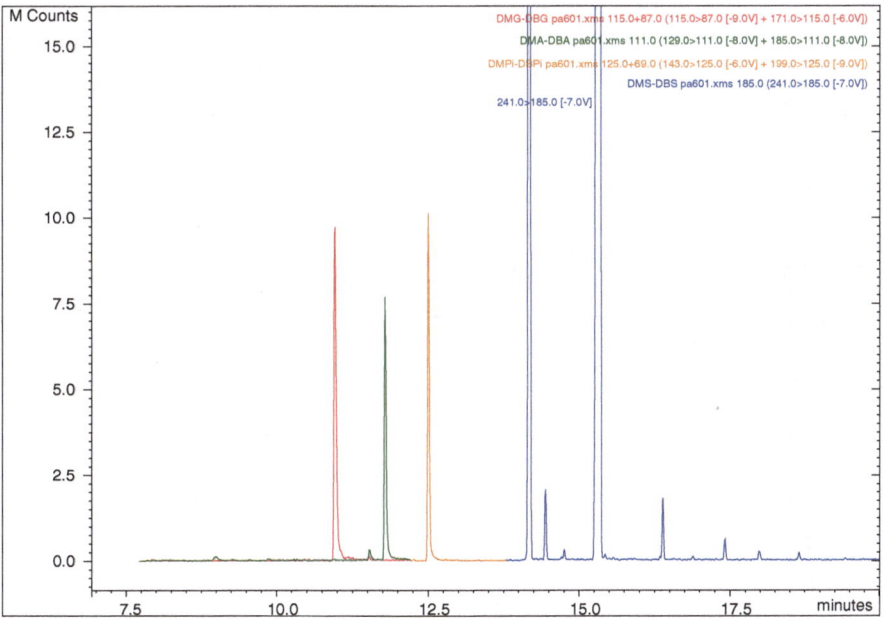

Fig. 3.13 Determination of polyadipates. GC/MS chromatogram of DMG (*red line*), DBA (*green line*), DBPi (*orange line*) and DBS (*blue line*) in oil

This procedure includes the use of internal standards: these substances are immediately added after sample weighing and before its dissolution in THF. Dimethyl pimelate (DMPi) is effectively used as an internal standard in the method of quantification of polyadipate plasticizer: the behaviour in the reaction of transesterification (times and percentage of reactions) is very similar to polyadipates [24].

Dimethyl glutarate (DMG) and dimethyl sebacate (DMS) are instead used as check standards. DMG has faster times of transesterification if compared with polyadipates and DMPi, while DMS has the slower times. However, 60 s of reaction should be sufficient for all involved substances [24]; therefore, DMG and DMS should be indicators of a good trend in the reaction.

At this stage, however, the obtained result (as dimethyl adipate) needs to be converted in polyadipate using a multiplying conversion factor. Probably, the calculation of this conversion factor is the most complex part of the analysis. As a consequence, polyadipate components (used as plasticiser in food packaging materials) with low molecular weights (MW) <1,000 Da must be preliminarily identified.

The chemical identification is carried out by GC/MS analysis of a solution of the polyadipate (if available) upon silylation reaction [25]. Because of the absence of polyadipate used as plasticiser, the original method [24] requires the solubilisation of a part of the packaging itself in THF.

After PVC precipitation (with methanol or ethanol) and the silylation reaction, the GC/MS analysis is carried out. The quantitative compositional analysis of low MW components is performed by means of a GC/FID apparatus with the same solution (as silylated polyadipate). Figure 3.14 shows GC/FID chromatograms of

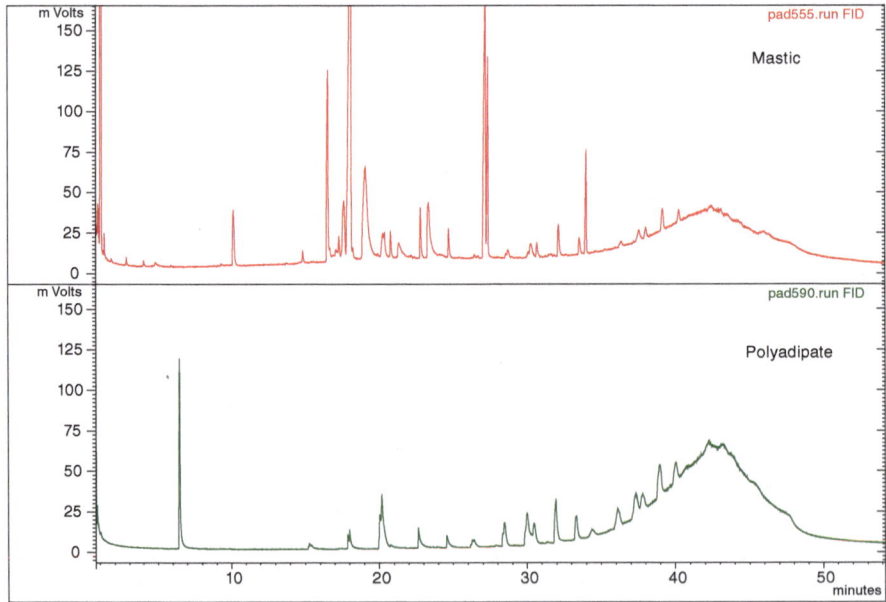

Fig. 3.14 GC/FID chromatograms of one pure polyadipate (*green line*) and the same obtained from the dissolution of mastic caps (*red line*)

a pure polyadipate and of the same set with other plasticisers obtained from the dissolution of mastic in capsules.

As a consequence, the useful conversion factor for the determination of polyadipates in food samples can be determined on the basis of GC/MS and GC/FID analyses [24]. The analysis of the profile obtained from pure polyadipate is certainly easier. At the same time, the profile obtained from the dissolution of plasticised PVC can give additional information concerning the presence of other plasticising substances.

3.3 Conclusions

PVC will probably be produced and still used for a long time, even if the criticism concerning the use of this material has taken place for several years. This problem concerns especially the production that involves highly toxic substances such as vinyl chloride and both waste disposal (production of dioxin during incineration, etc.).

The use of PVC for food packaging has not suffered further reductions (until now). However, several restrictions have already been taken and other limitation may be easily predicted in future, including the hypothesis of a replacement of PVC. Actually, dedicated studies on other materials and the evaluation of already existing materials for the replacement of PVC in food contact are being carried out. However, we can only try to use that material in the best way at the moment considering the knowledge and experience which have been gained so far.

Surveys and controls for packed food in glass jars and sealed with metal caps have been carried out by SSICA for several years [26, 27]. Research has mainly focused on oily foods or the contact of the capsules with the correct quantity of D simulant for these foods.

Results have shown that the traditional test based on regulations in force to evaluate new capsules before the contact with the food can hardly ever be representatives of what happens in the whole commercial life of the products. The development of a new specific test which could give a reliable forecast, before the contact, of migration before expiration and especially towards the end of the commercial life has failed. Probably, reasons for this situation are related to the remarkable diversification of food products and of heat treatments. Consequently, the type of product and the used technology can influence the commercial life of each product. It may be affirmed that the reproduction of standardised and short-lasting tests is difficult.

However, processes and analytical researches have definitely improved if compared to 10 years ago, when the first problems of specific migration in oily foods were tackled [28]. Nevertheless, it should be expected that the most efficient controls are carried out on packaged foods instead of simulating liquids. This approach implies obviously that related tests have to be carried out during the whole commercial life of food products. On these bases, the analytical reliability of obtained results—SML and global migration values up to the expiration date—can be confirmed [28]. A remarkable analytical work is clearly needed. On the other side, the above-mentioned approach appears the only reliable way for evaluating the trend and shelf life values of packaged foods from the viewpoint of chemical contamination.

References

1. EFSA (2005) Opinion of the scientific panel on food additives, flavourings, processing aids and materials in contact with food (AFC) on a request from the commission related to Bis(2-ethylhexyl)phthalate (DEHP) for use in food contact materials. EFSA J 243:1–20. doi:10.2903/j.efsa.2005.243
2. European Parliament and Council (2005) Directive 2005/84/EC of the European Parliament and of the Council of 14 Dec 2005 amending for the 22nd time Council Directive 76/769/EEC on the approximation of the laws, regulations and administrative provisions of the member states relating to restrictions on the marketing and use of certain dangerous substances and preparations (phthalates in toys and childcare articles). Off J Eur Union L344:40–43
3. Commission European (2011) Commission Regulation (EU) No. 10/2011 of 14 Jan 2011 on plastic materials and articles intended to come into contact with food. Off J Eur Union L12:1–88
4. EFSA (2005) Opinion of the scientific panel on food additives, flavouring, processing aids and materials in contact with food (AFC) on a request from the commission related to butylbenzylphthalate (BBP) for use in food contact materials. EFSA J 241:1–14. doi:10.2903/j.efsa.2005.241
5. EFSA (2005) Opinion of the scientific panel on food additives, flavouring, processing aids and materials in contact with food (AFC) on a request from the commission related to di-butylphthalate (DBP) for use in food contact materials. EFSA J 242:1–17. doi:10.2903/j.efsa.2005.242
6. EFSA (2005) Opinion of the scientific panel on food additives, flavouring, processing aids and materials in contact with food (AFC) on a request from the commission related to di-isodecylphthalate (DIDP) for use in food contact materials. EFSA J 245:1–14. doi:10.2903/j.efsa.2005.245
7. EFSA (2005) Opinion of the scientific panel on food additives, flavouring, processing aids and materials in contact with food (AFC) on a request from the commission related to di-isononylphthalate (DINP) for use in food contact materials. EFSA J 244:1–18. doi:10.2903/j.efsa.2005.244
8. Fantoni L, Simoneau C (2003) European survey of contamination of homogenized baby food by epoxidized soybean oil migration from plasticized PVC gaskets. Food Addit Contam 20:1087–1096. doi:10.1080/0265203031001615186
9. Fankauser Noti A, Fiselier K, Biedermann S, Grob K, Armellini F, Rieger K, Skjevrak I (2005) Epoxidized soybean oil (ESBO) migrating from the gaskets of lids into food packed in glass jars. Eur Food Res Technol 221:416–442. doi:10.1007/s00217-005-1194-4
10. European Commission (2007) Commission Regulation No. 372/2007 laying down transitional migration limits for plasticisers in gaskets in lids intended to come into contact with foods. Off J Eur Union L92:9–12. Amended by European Commission (2008) Commission Regulation No. 597/2008 Off J Eur Union L164:12–13 on transitional migration limits for plasticisers in gasket in lids intended to come into contact with foods
11. Welle F, Wolz G, Franz R (2004) Study on the migration behaviour of DEHP versus an alternative plasticiser, Hexamoll® DINCH, from PVC tubes into enteral feeding solutions. Poster presented at the 3rd international symposium on food packaging, Barcelona, 17–19 Nov 2004

12. Petersen JH, Brendahl T (2000) Plasticizers in total diet samples, baby food and infant formulae. Food Addit Contam 17:133–141. doi:10.1080/026520300283487
13. Fankhauser-Noti A, Grob K (2006) Injector-internal thermal desorption from edible oils performed by programmed temperature vaporizing (PTV) injection. J Sep Sci 29:2365–2374. doi:10.1002/jssc.200600064
14. Tsumura Y, Ishimitsu S, Saito I, Sakai H, Tsuchida Y, Tonogai Y (2003) Estimated daily intake of plasticizers in 1-week duplicate diet samples following regulation of DEHP-containing PVC gloves in Japan. Food Addit Contam 20(4):317–324. doi:10.1080/0265203031000122021
15. Fusari P, Rovellini P (2009) Liquid chromatography-Ion Trap-ESI-mass spectrometry in food safety assessment: phthalates in vegetable oils. Riv Ital Sostanze Grasse 86(1):25–30
16. Dirwono W, Nam YS, Park HM, Lee KB (2013) LC-TOF/MS determination of phthalates in ediblesalts from food markets in Korea. Food Addit Contam B Surveill 6(3):203–208. doi:10.1080/19393210.2013.795194
17. Sannino A (2010) Development of a gas chromatographic/mass spectrometric method for determination of phthalates in oily foods. J AOAC Int 93(1):315–322
18. Technical Committee CEN/TC 275 (2008) EN 15662:2008. Foods of plant origin—determination of pesticide residues using GC-MS and/or LC-MS/MS following acetonitrile extraction/partitioning and clean-up by dispersive SPE—QuEChERS-method. European Committee for Standardization, Brussels
19. Castle L, Sharman M, Gilbert J (1988) Gas chromatographic-mass spectrometric determination of epoxidized soybean oil contamination of foods by migration from plastic packaging. J Assoc Off Anal Chem 71(6):1183–1186
20. Fankhauser-Noti A, Fiselier K, Biedermann-Brem S, Grob K (2005) Epoxidized soy bean oil migrating from the gaskets of lids into food packed in glass jars: analysis by on-line liquid chromatography-gas chromatography. J Chromatogr A 1082:214–218. doi:10.1016/j.chroma.2005.05.057
21. Biedermann-Brem S, Biedermann M, Fankhauser-Noti A, Grob K, Helling R (2007) Determination of epoxidized soy bean oil (ESBO) in oily foods by GC–FID or GC–MS analysis of the methyl diepoxy linoleate. Eur Food Res Technol 224(3):309–314. doi:10.1007/s00217-006-0424-8
22. Suman M, La Tegola S, Catellani D, Bersellini U (2005) Liquid chromatography-electrospray ionization-tandem mass spectrometry method for the determination of epoxidized soybean oil in food products. J Agric Food Chem 53(26):9879–9884. doi:10.1021/jf052151x
23. Castle L, Mercer AJ, Gilbert J (1988) Gas chromatographic-mass spectrometric determination of adipate-based polymeric plasticizers in foods. J Assoc Off Anal Chem 71(2):394–396
24. Biedermann M, Grob K (2006) GC Method for determining polyadipate plasticizers in foods: transesterification to dibutyl adipate, conversion to migrating polyadipate. Chromatographia 64:543–552. doi:10.1365/s10337-006-0071-z
25. Biederman M, Grob K (2006) GC–MS characterization of oligomers in polyadipates used as plasticizers for PVC in food contact. Packag Technol Sci 19:159–178. doi:10.1002/pts.722
26. Graubardt N, Biedermann M, Fiselier K, Bolzoni L, Pedrelli T, Cavalieri C, Simoneau C, Grob K (2009) Search for a more adequate test to predict the long-term migration from the PVC gaskets of metal lids into oily foods in glass jars. Food Addit Contam A 26(7):1113–1122. doi:10.1080/02652030902894405
27. Graubardt N, Biedermann M, Fiselier K, Fiselier L, Bolzoni L, Cavalieri C, Grob K (2009) Further insights into the mechanism of migration from PVC gaskets of metal closures into oily foods in glass jars. Food Addit Contam A 26(8):1217–1225. doi:10.1080/02652030902950835
28. McCombie G, Harling-Vollmer A, Morandini M, Schmäschke G, Pechstein S, Altkofer W, Biedermann M, Biedermann-Brem S, Zurfluh M, Sutter G, Landis M, Grob K (2012) Migration of plasticizers from the gaskets of lids into oily food in glass jars: a European enforcement campaign. Eur Food Res Technol 235(1):129–137. doi:10.1007/s00217-012-1739-2

Chapter 4
Organic Food Packaging Contaminants: New and Emerging Risks

Salvatore Parisi, Caterina Barone and Giorgia Caruso

Abstract By the chemical viewpoint, the most part of food contact-approved materials is composed of organic molecules. Consequently, the chemical classification of food packaging products should be carefully evaluated because of two different reasons at least: the technological process (or sum of subprocesses) and the final structure (or multi-layered system) of the food container, when applicable. In fact, chemical features of raw materials and different intermediates influence the peculiar process of productions in the packaging sector. As a result, many possibilities—excellent or reliable packaging materials—may be early discarded in the initial design stages because of practical factors: the packaging line may appear unsuitable for the peculiar container, the processed food needs more resistant materials during and after the packing step and so on. These situations and other possible reasons suggest caution. Consequently, many risks can be avoided by the microbiological or physical viewpoint, according to the 'hazard analysis and critical control points' (HACCP) approach. However, chemical contaminants may occur. This work would discuss most part of the known organic contaminants in food products on the basis of past experiences and risk assessment approaches in the food industry.

Keywords Chemical contaminant · Downstream user · Food additive · Food hygiene · Food packaging material · REACH

Abbreviations

DEHP	Bis(2-ethylhexyl) phthalate
BPA	Bisphenol A
BADGE	Bisphenol A diglycidyl ether
BFDGE	Bisphenol F diglycidyl ether
BRC	British Retail Consortium
BSI	British Standards Institution
CAS	Chemical Abstract Service

© The Author(s) 2015
C. Barone et al., *Food Packaging Hygiene*, Chemistry of Foods,
DOI 10.1007/978-3-319-14827-4_4

DPB	Dibutyl phthalate
DIBP	Diisobutyl phthalate
DIPN	Diisopropyl naphthalene
DPP	Dipentyl phthalate
DU	Downstream user
EMA	Economically motivated adulteration
EDC	Endocrine disruptor compound
EFSA	European Food Safety Authority
EU	European Union
FRF	Fat consumption reduction factor
FWA	Fluorescent whitening agent
FPM	Food packaging material
FPP	Food packaging producer
FP	Food producer
FQMS	Food quality management system
GSFS	Global Standard for Food Safety
GMP	Good manufacturing practices
HACCP	Hazard analysis and critical control points
IoP	Institute of Packaging
IFP	Integrated food product
IFS	International Featured Standards
ITX	2-isopropylthioxantone
MOSH	Mineral oil saturated hydrocarbon
MOAH	Mineral oil aromatic hydrocarbon
MFFB	Moisture on free fat basis
NIAS	Non-intentionally added substance
NOGE	Novolac glycidyl ether
NCFPD	National Center for Food Protection and Defense
OML	Overall migration limit
P&B	Paper and board
PCP	2,3,4,5,6-pentachlorophenol
POP	Persistent organic pollutant
PCB	Polychlorobyphenyl
PAH	Polycyclic aromatic hydrocarbon
PAA	Primary aromatic amine
PAS	Publicly Available Standard
PVC	Polyvinyl chloride
REACH	Registration, Evaluation, Authorisation and Restriction of Chemicals
SVOC	Semi-volatile organic compound
SML	Specific migration limit
SVHC	Substance of very high concern
TiO_2	Titanium dioxide
UV	Ultraviolet
USP	United States Pharmacopeia
VOC	Volatile organic compound

4.1 The Food Industry and the HACCP Approach

The approach of the food industry to safety hazards has been deeply reviewed since 1990s, from the simple management of purchased raw materials with the 'first-in-first-out' strategy or different methods to the preventive evaluation of hygiene conditions and correlated risks [1, 2]. In fact, the main target of the food producer (FB), packer or retailer is the satisfaction of the final consumer. Consequently, every food-related hazard has to be seen from the viewpoint of this subject. From the regulatory viewpoint, the FB is entirely responsible for its own product, including possible damages for the human health.

The concept of food safety is strictly related to the definition of three different hazards (Sect. 1.1):

1. The microbiological risk, with reference to every pathogen or dangerous bio-logical agent, including possible associations between similar and different life forms. Actually, other non-biological hazards may be considered in relation to the microbiological risk
2. The chemical risk, with reference to every chemical dangerous substance. Actually, this concept cannot be easily circumscribed: different categories of dangerous chemicals have been classified with several 'empty spaces', and new emergencies may occur at present. One of the most recent worries concerns nanomaterials
3. The physical risk, with specific reference to every typology of foreign and dangerous macroscopic materials into packed foods and beverages. Basically, the physical risk can be evaluated when the presence of wooden particles, steel parts, glass materials, etc., may be demonstrated or supposed into foods, with consequent and immediate damages to the human health by simple ingestion. It should be noted that this topic is circumscribed enough, but different strategies can be used for the eradication or the reduction of the physical risk.

The evaluation of these risks is one of main pilasters of the 'hazard analysis and critical control points' (HACCP) approach. According to basic principles of HACCP, every possible risk has to be studied with an introductory analysis of so-called nonconformities [2]. The analysis of studies and the statistical evaluation of nonconformities can be discussed in the food industry in the ambit of the HACCP team: this structure means a specific group composed of different but necessary key functions, including the legal direction (chairman, chief executive director, etc.) and several roles for main departments: planning, production, purchasing, warehouse, maintenance and laboratory (also named quality control). Additionally, one or more external consultants may be needed. Anyway, the HACCP group has to be chaired by the HACCP manager.

Finally, the result of studies, discussions and the preliminary implementation of corrective actions has to be formally written and recorded: in other words, the HACCP plan has to be declared and subsequently applied and demonstrable.

The creation and the final implementation of the HACCP plan do not exclude subsequent reviews and modifications when new possible risks occur or the regulatory situation is modified: basically, the HACCP plan is 'live' and continuously revalidated.

This chapter concerns the problem of the chemical contamination in foods by food packaging materials (FPM). As a clear consequence, microbiological and physical hazards should not be discussed. However, many menaces to food safety appear to be caused by microbial spreading: the same thing might be inferred when speaking of simple contamination by FPM [3]. At first sight, the presence of active life forms into foods and/or edible raw materials for the production of packed foods can surely cause (or be correlated with) different chemical risks. Actually, the same situation can occur when these micro-organisms are found on processing or packing equipment [4].

The synthesis of various enterotoxins by *Escherichia coli* and other life forms (e.g.: *Clostridium botulinum* and *Clostridium perfringens*) is one of most known worries by the hygiene viewpoint. According to different authors [5–7], the ingestion of similar toxins causes pathological phenomena with the possible death of human consumers. Consequently, the risk has to be attentively monitored, and adequate preventive and corrective actions have to be put in place: basically, the microbiological aspect seems the main and common feature in these situations. However, the evaluation of risks should be made on the basis of a statistical study, and the analytical detection of different toxins is mainly correlated to the work of research chemists. In fact, there is not a sure correlation between live micro-organisms and the presence of thermostable molecules with dangerous effects: on the contrary, chemically contaminated foods might be found with inactivated bacteria after thermal treatments or other preservation techniques. Because of the presence of residual toxins, the microbiological surveillance is extremely important, but the second step remains the analytical control of toxins.

Secondly, different microbial groups can also produce molecules and substances with non-toxic or harmful effect on human health. However, these compounds may cause unacceptable degradations on packed foods: the importance of sensorial evaluations [8] should be highlighted. In effect, the correct evaluation of colours, odours, texture and other simple parameters can give important and useful information with concern to the occurrence of microbial spreading and the possible modification of main chemical data. For example, the direct correlation between the softness and the moisture on free fat basis (MFFB) index in several 'pasta filata' soft cheeses is well known [9, 10]. It should be considered also that the microbial spreading causes appreciable augments of the aqueous content in these products because of hydrolysis.

Generally, all possible food alterations caused by microbial spreading appear connected to several known sources: raw materials, environmental conditions, poor sanitation, etc. The role of FPM is not mentioned. However, microbial spreading and physical–chemical features of FPM can synergically act. Should this situation be verified, the 'technological suitability' of the peculiar container or FPM component would not be sure. Following factors should be carefully examined:

- Incorrect design of FPM
- Defective communication between FP and food packaging producers (FPP)
- Failures of the FPM production process (assembling steps are also included here)
- Other secondary causes.

Anyway, every possible failure of the so-called integrated food product (IFP) caused by microbial activity may be potentially increased if the used FPM is not suitable for the specific use [4, 11]. According to Sect. 1.1, IFP means the association of FPM and the packed food with other visible and immaterial components.

The perception of FPM as an accessory or secondary component of the IFP has been already discussed in Sect. 1.1. In reference to this approach, all parts of the IFP can be potentially damaged with microscopic or macroscopic effects; sometimes, the exact adjective could be 'grotesque'. As a result, both food content and FPM can suffer light or high failures at the same time because of microbial spreading: this simple deduction can be useful with concern to the importance of FPM as a primary component of the IFP, in spite of the non-edible nature.

With concern to the possible presence of foreign and dangerous materials in packed foods, many reflections can be made, but this book is not specifically dedicated to the physical risk. Certainly, the detection of metallic powders or similar objects can be caused by contact interactions between food and FPM, but the most part of macroscopic residues is generally correlated with anomalies on processing and/or packing lines and related equipment. The same thing may be inferred when speaking of microbiological contamination: the presence of biofilms and 'protobiofilms' (small colonies of micro-organisms during food processes) on the surface of processing machinery is extremely important [2, 12–14].

With exclusive reference to the chemical contamination by FPM contact without microbial spreading or migration of macroscopic substances, following substances should be mentioned:

- Forbidden additives
- Chemical substances with hygienic concerns, limitations or peculiar obligations (labelling). Examples: tartrazine, E102; quinoline yellow, E104
- Food allergens or related categories
- Genetically modified organisms
- Irradiated foods or raw materials
- Radioactive elements. Examples: caesium and strontium
- Mycotoxins. Examples: aflatoxin M1
- Heavy metals (example: lead), dioxins and dioxin-like compounds and polychlorobyphenyls (PCB)
- Pesticides
- Antibiotics and hormone residues. Examples: chloramphenicol, chloroform, chlorpromazine, colchicine, metronidazole, nitrofuran and furazolidone.

Naturally, this list is not exhaustive: for example, one of the main discussions concerns the presence of nanomaterials in foods at present.

- In relation to all above-mentioned chemicals, it has to be clarified that the origin of contamination is often searched (and found) within the production cycle of the packed food with the exclusion of the packaging step. This approach is reliable when speaking of normal food additives. On the other hand, several contaminants may be found on different FPM or food contact-approved surfaces. Two examples can be given as follows [4, 13, 15, 16]:
  - Plastic moulds for the production of cheese intermediates in the dairy industry
  - Self-lubricating coatings for food processing lines.

The problem of chemical contamination in foods and beverages by different sources is a very interesting argument, but the detailed discussion would need many pages. In reference to basic aims of this book, the declared intention is to provide a sort of introductory overview of organic contaminants from FPM. Actually, there is a notable part of inorganic chemicals, but these food contaminants are discussed in Chap. 2.

Organic chemical residues are certainly a well-known risk in the food industry, in accordance with the current legislation in different countries; most known food quality standards have already considered the problem. However, the regulatory aspect has been previously discussed in this book from a general viewpoint (Sect. 1.2). Section 4.2 is dedicated to most known organic contaminants in packed foods by FPM.

4.2 Known Chemical Risks in the Food Industry and the Connection with Food Packaging Materials

Basically, every food or beverage product has its own chemical composition. This feature depends on several factors: two of these variables are the type of raw materials and the design of the food technologist.

The possible addition of chemical substances, also named 'food additives', is surely important: according to the 'Codex General Standard for Food Additives', clause 3.2 [17], these compounds can be justified in the formulation of food products if

- Their presence represents an advantage
- The human health is not damaged
- Their presence is not cause of prevarication for the consumer. In other words, the food products do not reproduce distinguished and known features of other foods with a clear fraudulent intent
- Their use can help food technologists to obtain important technological functions
- The evidence of above-mentioned enhancements could not be achieved by means of different systems with an economic and technological advantage.

These points are extremely important: in fact, the improper use of food additives may potentially generate food safety dangers. With explicit reference to the clause 3.2 of the above-mentioned 'Codex General Standard for Food Additives', following concerns may occur by the exclusive viewpoint of research chemists [17]:

1. Basically, the addition of a peculiar food additive may alter the nutritional composition of foods, and the relationship between edible products and FPM may suffer important modifications. These alterations should be preventively investigated [4, 13]
2. The perception of sensorial features of foods and beverages may be altered (cause: incorrect addition of food additives)
3. The fraudulent addition of several chemicals remains to be discussed. The recent 'Economically Motivated Adulteration' (EMA) project, promoted by the National Center for Food Protection and Defense (NCFPD) and the United States Pharmacopeia (USP), demonstrates more research is still needed about this 'thorny' topic [18].

With reference to the possible influence of FPM, it has to be noted that

- Several food additives may show notable similarity with common components of FPM. For example, the inorganic pigment titanium dioxide (TiO_2) is mentioned in the list of allowed additives of the 'Codex General Standard for Food Additives' [17]; it is also defined as E171. However, this substance is largely used in the production of white enamels for metal cans and other industrial non-edible products
- The interaction between foods and FPM at the contact interface may give use some surprise from the organoleptic viewpoint. These modifications of the edible content may have other causes also (operational conditions of food processing, storage protocols, etc.), but their visible effect may often change depending on the peculiar formulation of the edible food.

As a result, words 'chemical contamination' may mean a number of situations and mechanisms of transfer and/or chemical reactions. By the HACCP viewpoint, apparent or real chemical hazards are evident when the original (unpacked) food is modified in comparison with the designed composition because of

1. Diffusion of foreign but edible contaminants in the inner and/or external layers, including the superficial area
2. Diffusion of foreign and non-edible contaminants in the inner and/or external layers, including the superficial area
3. Transformation of one or more original components of the final IFP because of predictable or unknown factors, with active influence of FPM
4. Transformation of one or more original components of the final IFP because of predictable or unknown factors under incorrect storage conditions, without active influence of FPM

5. Transformation of one or more original components of the final IFP because of predictable or unknown factors under incorrect storage conditions, with active influence of FPM
6. Apparent transformation of sensorial features because of predictable or unknown factors under normal or incorrect storage conditions, with or without FPM ruptures or other damages.

Actually, all above-mentioned phenomena can concern food contamination by organic and inorganic substances at the same time. This discussion is related to organic contaminants only. In relation to these compounds, it can be preliminary affirmed that most common organic contaminants by FPM are generally [4, 19]:

(a) Composite materials with predominant presence of plastic molecules and related polymerisation intermediates. These materials may release organic molecules with good or acceptable solubility in organic solvents and low solubility in water. As a consequence, released chemicals may be found in fat and medium-fat foods. Moreover, dry foods might show limited amounts of these contaminants because of the chemical similarity with several organic components

(b) Composite materials with cellulosic matters and plastic coatings (polycoupled packages). These materials may release organic molecules with good or acceptable solubility in organic solvents and low solubility in water. As a consequence, released chemicals may be found in fat and medium-fat foods. In addition, several dry foods have been found with limited amounts of these contaminants because of the chemical similarity with organic components. However, different causes should be evaluated [19]

(c) Metallic materials (metal cans) with the protection of plastic coatings and enamels. These materials may release organic molecules with good or acceptable solubility in organic solvents and low solubility in water because of plastic coatings or enamels. As a consequence, released chemicals may be found in fat and medium-fat foods. Moreover, dry foods might show limited amounts of these contaminants because of the chemical similarity with organic components and storage conditions. Normally, room temperature is considered adequate for the preservation of canned foods until the end of shelf life

(d) Composite materials—plastic-made containers, paper and board, polycoupled packages, metal cans, etc.—by recycled materials.

Except for glass containers and FPM with low diffusion in the world of food distribution, the position of plastic materials and related intermediates is not excluded when speaking of chemical alterations. As a clear result, the contamination of foods is presumptively connected to the presence of peculiar organic chemicals. On these bases, several of the most important organic food contaminants are often highlighted as FPM-related menaces. Section 4.3 is dedicated to several or these molecules or groups of compounds by the European viewpoint.

4.3 Most Known Organic Contaminants in Food Packaging Materials: The European Viewpoint

Different approaches can be observed in various countries, but several similarities may be also found at present. With reference to our discussion, the European vision of FPM may be shown as an useful example.

First of all, the Regulation (EC) No 1935/2004 states clearly that every potential FPM contaminant has to be correctly identified. From a general viewpoint, this requisite is needed when speaking of all possible FPM from virgin sources, recycled materials and mixed raw materials. Actually, the situation may appear too specific and difficult with reference to concrete measures for FPM. However, several national and international institutions have already faced the problem with the consequent publication of dedicated guidelines. For example, the Italian Institute of Packaging has recently issued its guideline about the evaluation of the Declaration of Food Contact Compliance for FPM obtained by recycled raw materials [20].

Moreover, FPM should be subdivided into different groups, depending on their nature and the final destination at least, as shown in the recent literature [4]. Generally, this subdivision may be performed in the following manner:

- Plastic containers
- Paper and board (P&B)
- Glass FPM
- Metal cans.

With reference to most known food contaminants, it should also recognised they may be grouped in a relatively short and 'transversal' list. In effect, a notable part of these compounds may migrate from different FPM. However, a sort of simplification may be operated.

First of all, metals and inorganic compounds may be eliminated from the present discussion; Chap. 2 is dedicated to this group of contaminants. These dangers are often linked to the composition of metal containers and glass FPM.

Secondly, plastic and P&B sectors appear strictly connected when speaking of polycoupled packages [4]. This connection is probably appreciable with concern to technological issues and 'shared' contaminants. Consequently, every dangerous chemical substance related to the industry of plastic matters may be found or originated by both plastic and P&B containers. Finally, it should be remembered metal cans are hybrid (plastic/metallic) packaging materials because of the presence of plastic coatings and enamels on the inner food contact side.

As a result, one of the above-mentioned FPM sectors can be explored with the aim of discussing most organic contaminants. The field of P&B packages is interesting because of the clear connection with polycoupled FPM and the use of recycled raw materials. At present, more of 50 % of the whole market of P&B PFM is constituted of recycled fibres [20, 21].

Generally, main organic contaminants in the field of P&B materials are listed as follows [20]:

- 2,3,4,5,6-pentachlorophenol (PCP)
- Phthalates
- Volatile organic compounds (VOC) and semi-volatile organic compounds (SVOC)
- Diisopropylnaphthalene (DIPN)
- Polycyclic aromatic hydrocarbons (PAH)
- Formaldehyde and glioxal
- Polychlorobyphenyls (PCB)
- Primary aromatic amines (PAA)
- Fluorescent whitening agents (FWA)
- Photoinitiators
- Bisphenol A (BPA)
- Mineral oil saturated hydrocarbons (MOSH) and mineral oil aromatic hydrocarbons (MOAH)
- Microbiological agents: yeasts and moulds.

Above-mentioned contaminants may be summarised in the following manner.

PCP, chemical formula: C_6HCl_5O, CAS Number: 87-86-5 is an excellent biocide against the action of moulds.

Phthalates are used as plastifiers in the production of polyvinyl chloride (PVC). Three of these compounds are as follows:

- Dibutyl phthalate (DPB), chemical formula: $C_{16}H_{22}O_4$, CAS number: 84-74-2
- Bis(2-ethylhexyl) phthalate (DEHP), chemical formula: $C_{24}H_{38}O_4$, CAS number: 117-81-7
- Diisobutyl phthalate (DIBP), chemical formula: $C_{16}H_{20}O_4$, CAS number: 84-69-5.

It has to be noted that the whole class of phthalates comprehends 'endocrine disruptor compounds' (EDC): these compounds are also defined 'persistent organic pollutants' (POPs) because of the well-known duration and bioaccumulation in the environment [20]. With reference to analytical methods, the prEN 16453: 2012 norm (pulp, paper and paperboard—determination of phthalates in extracts from paper and paperboard) is dedicated to the detection of DPB, DEHP and DIBP at least.

Organic solvents for inks and coatings comprehend a number of different substances, including VOC. One of these compounds is the well-known ethyl acetate.

DIPN, a mixture of isomeric diisopropylnaphthylenes, is not used in the production of FPM; consequently, it should be defined 'non-intentionally added substance '(NIAS).

PAH are found in ultraviolet (UV) inks and photoinitiators for UV inks, but their presence does not imply peculiar functions. Substantially, PAH should be considered simple contaminants when speaking of ink formulations.

Formaldehyde, chemical formula: CH_2O, CAS number: 50-00-0. It is extensively used for the formulation of glues; moreover, formaldehyde can enhance the resistance of several resins to moisture. The same property is shown by glioxal, chemical formula: $C_2H_2O_2$, CAS number: 107-22-2.

PCB are not allowed for the production of copying paper in the EU. Consequently, their detection should be possible on recycled materials only. In effect, the research of PCB is not compulsory in several EU countries at present.

PAA are associated to azo dyes, but their residual presence is caused by incomplete polymerisation or decomposition. Anyway, PAA are not used as food colourants.

FWA are added to paper materials with the aim of enhancing UV radiations. Because of the shift of the emitted light from papers under exposition to sunlight, added materials show the well-know fluorescence with the resulting increase of white tonalities. FWA are not allowed for the production of food contact PFM; their presence into recycled fibres cannot be permitted.

Photoinitiators are used for the reticulation of UV coatings and inks under UV light sources [11]. For examples:

1. Benzophenone, chemical formula: $C_{16}H_{20}O_4$, CAS number: 119-61-9
2. 4,4′-bis(dimethylamino) benzophenone, also named Michler's ketone, chemical formula: $C_{17}H_{20}N_2O$, CAS Number: 90-94-8.

BPA, chemical formula: $C_{15}H_{16}O_2$, CAS Number: 80-05-7, is a monomer for the production of epoxidic resins [11]; it can be added to PVC articles. Other uses are well known in the industry of plastic materials for non-food contact applications. Recently, BPA has been banned in different countries without a clear and harmonised action [4].

With concern to MOSH and MOAH, the toxicological status of these substances is not clear [19]: the European Food Safety Authority (EFSA) and other national agencies in the EU do not consider MOSH and MOAH safety concerns, at present. However, the limitation of these contaminants in FPM has been recently discussed in the EU [21] because MOSH and MOAH are surely undesirable in food products. On the other hand, mineral oils are detectable in a number of industrial (edible and non-edible) products. The detection of mineral oils in dry foods has been recently demonstrated as caused by secondary packages in spite of the nature of 'barrier' of the primary package (polyethylene plastic bags) [19, 21].

Anyway, the food sector is accustomed to following mineral oils [17, 19, 21]:

- Mineral oil, medium and low viscosity, Class I (EU classification: E905e); use: glazing agent
- Mineral oil, high viscosity (EU classification: E905d); use: antifoaming agent, glazing agent
- Microcrystalline wax (EU classification: E905c). Possible uses: antifoaming agent, glazing agent, coating agent for cheeses (this wax can efficaciously avoid the migration of moisture out of the food product).

The use of MOSH and MOAH in the EU is not restricted at present: in other words, should these mineral oils be detected in foods and beverages, related products would not be subjected to peculiar sanctions. Every recall and/or withdrawal procedures should be justified as a preventive measure for the safeguard of consumers by the hygiene viewpoint.

Finally, yeasts and moulds are common life forms: these organisms are researched in the most part of P&B containers for food applications.

4.4 Other Problems: Substances of Very High Concern

The discussion about food contaminants by FPM in the EU should be concluded with the recent (EC) Regulation No 1907/2006 concerning the Registration, Evaluation, Authorisation and Restriction of chemicals (REACH). Actually, this document is not specifically related to food contaminants, the world of the food production and FPM. However, the importance of REACH cannot be underestimated because of the new procedure of registration and evaluation of chemical substances in the EU. As a consequence, the use of chemicals may be authorised or prohibited; actually, possible restrictions are allowed if necessary. Anyway, the main target of the REACH is the safeguard of the human health and the environment.

By a general viewpoint, the identification and the consequent evaluation of chemical substances can be proposed by the European Commission or one of the EU member countries. Subsequently, the use of these chemical compounds can be authorised, prohibited or restricted.

With concern to the present discussion about food contaminants, every FP is considered 'downstream user' (DU) according to REACH; this concept is applicable to FPP also because of the use of a number of different chemicals or chemical mixtures. By the REACH viewpoint, DU is obliged to declare other possible uses of chemical additives if they are different from the most known applications or recommendations. On these bases, existing substances may be authorised for other uses, or a detailed prohibition (or restriction) may be decided.

Basically, the aim of general REACH procedures is the inclusion of every examined substance in one of two possible lists. The first of these documents, undoubtedly the most important, is the 'Authorisation List': it comprehends only 'substances of very high concern' (SVHC) because these compounds have been recognised at least:

- Carcinogenic, and/or
- Mutagenic, and/or
- Toxic for reproduction, and/or
- Very persistent and bioaccumulative, or persistent, bioaccumulative and toxic according to the REACH Regulation, Annexes XII and XIII.

Consequently, SVHC are prohibited in the EU except for possible and specific exemptions. Actually, the inclusion of a peculiar chemical in the Authorisation

List does not imply automatically specific migration limits (SML) with concern to FPM; on the other hand, the detection of SVHC into foods is surely a big concern.

The Authorisation List is constantly under revision and publicly available: at present, 22 different substances have been listed and classified SVHC in accordance with the Regulation (EU) No 348/2013 [22]. It should be noted that several of these compounds have been previously mentioned in Sect. 4.3. For example, following phthalates are considered SVHC:

- Benzyl butyl phthalate (BBP), EC number: 201-622-7, CAS number: 85-68-7
- Bis(2-ethylhexyl) phthalate (DEHP), EC number: 204-211-0, CAS number: 117-81-7
- DBP, EC number: 201-557-4, CAS number: 84-74-2
- DIBP, EC number: 201-553-2, CAS number: 84-69-5.

Moreover, the first three of these phthalates may be used in the immediate packaging of medicinal products covered under Regulation (EC) No 726/2004, Directive 2001/82/EC and/or Directive 2001/83/EC. This is an example of specific exemption.

Consequently, it can be assumed the inclusion of a peculiar substance in the EU Authorisation List corresponds to a real 'alarm bell', and DU has to take note of this advice. For example, the position of mineral oils is unclear at present in the EU because these compounds are not currently classified SVHC.

Actually, REACH procedures can generate two different lists of substances: the second of these documents has to be mentioned also. The 'Candidate List of SVHC for Authorization' contains all substances with a proposal of inclusion in the Authorisation List. Should these chemicals be considered SVHC, they would be transferred in the Authorisation List.

This Candidate List is naturally under constant revision and publicly available: at present, 144 different substances have been listed [23]. Once more, there are not sure correlations between the possible inclusion of a peculiar substance in the Candidate List and the definition of specific SML. On the other hand, it has to be recognised that the simple mention of one of these 144 'nominated' substances by specialised media in relation to food scandals can surely pose a serious problem to food players. For example, the following list of organic substances are mentioned in the Candidate List:

- Dipentyl phthalate (DPP), EC number: 205-017-9, CAS number: 131-18-0
- o-Toluidine, EC number: 202-429-0, CAS number, 95-53-4
- N, N-dimethylformamide, EC number: 200-679-5, CAS number, 68-12-2
- N-pentyl-isopentylphthalate, CAS Number, 776297-69-9.

As a result, future EU food scandals may be easily correlated to one or more of mentioned chemicals in the Authorisation or the Candidate Lists. Because of the growing dimension of the second list, it may be anticipated that the number of possible food contaminants is destined to grow rapidly in the same way.

References

1. Ko WH (2013) The relationship among food safety knowledge, attitudes and self-reported HACCP practices in restaurant employees. Food Control 29:192–197. doi:10.1016/j.foodcont.2012.05.076
2. Ottaviani F (2002) Il metodo HACCP (hazard analysis and critical control points). In: Andreis G, Ottaviani F (eds) Manuale di sicurezza degli alimenti. Principi di ecologia microbica e di legislazione applicati alla produzione alimentare. Oxoid S.p.A., G. Milanese, Milan
3. Parisi S (2011) Food packaging and technological compliance the importance of correct storage procedures. Food Packag Bull 20(9 and 10):14–18
4. Parisi S (2012) Food packaging and food alterations: the user-oriented approach. Smithers Rapra Technology, Shawbury
5. Ottaviani M (2002) Muffe e Micotossine. In: Andreis G, Ottaviani F (eds) Manuale di sicurezza degli alimenti. Principi di ecologia microbica e di legislazione applicati alla produzione alimentare. Oxoid S.p.A., G. Milanese, Milan
6. Milićević D Rl, Škrinjar M, Baltić T (2010) Real and perceived risks for mycotoxin contamination in foods and feeds: challenges for food safety control. Toxins 2:572–592. doi:10.3390/toxins2040572
7. Karmali MA (2004) Infection by shiga toxin-producing *Escherichia coli*. Mol Biotechnol 26:117–122. doi:10.1385/MB:26:2:117
8. Gioffrè ME, Parisi S, Piccione D, Micali M, Delia S, Laganà P (2009). Raffronto tra analisi microbiologiche e valutazione organolettica degli alimenti. Casi di studio in diversi comparti alimentari. In Ig Sanità Pubbl 5/2009, supplement, p 392
9. Parisi S, Laganà P, Delia AS (2006) Il calcolo indiretto del tenore proteico nei formaggi: il metodo CYPEP. Ind Aliment 462:997–1010
10. Parisi S, Laganà P, Stilo A, Micali M, Piccione D, Delia S (2009) Il massimo assorbimento idrico nei formaggi. Tripartizione del contenuto acquoso per mole d'azoto. Ind Aliment 491:31–41
11. Parisi S (2004) Alterazioni in imballaggi metallici termicamente processati. Gulotta Press, Palermo
12. Parisi S (2010) HACCP assessment: studio di nuovi piani di campionamento in forma ridotta nel campo lattiero-caseario. Dissertation, University of Messina
13. Parisi S (2013) Food industry and packaging materials—performance-oriented guidelines for users. Smithers Rapra Technology, Shawbury
14. Poulsen LV (1999) Microbial biofilm in food processing. LWT—Food Sci Technol 32:321–326. doi:10.1006/fstl.1999.0561
15. Micali M, Parisi S, Minutoli E, Delia S, Laganà P (2009) Alimenti confezionati e atmosfera modificata. Caratteristiche basilari, nuove procedure, applicazioni pratiche. Ind Aliment 489:35–43
16. Lau OW, Wong SK (2000) Contamination in food from packaging material. J Chromatogr A 882:255–270. doi:10.1016/S0021-9673(00)00356-3
17. Codex Alimentarius Commission (1995) Codex general standard for food additives, last revision 2013. Codex alimentarius—international food standards. http://www.codexalimentarius.net/gsfaonline/docs/CXS_192e.pdf. Accessed 18 Oct 2013
18. Institute of Food Technologists (2013). University of Minnesota launches databases to prevent food adulteration. http://www.ift.org/food-technology/daily-news/2013/february/06/univ-of-minnesota-launches-databases-to-prevent-food-adulteration.aspx. Accessed 18 Oct 2013
19. Vollmer A, Biedermann M, Grundböck F, Ingenhoff J-E, Biedermann-Brem S, Altkofer W, Grob K (2011) Migration of mineral oil from printed paperboard into dry foods: survey of the German market. Eur Food Res Technol 232:175–182. doi:10.1007/s00217-010-1376-6
20. Italian Institute of Packaging (2013) Linee guida per la valutazione dell'idoneità al contatto con alimenti del packaging realizzato con materiale proveniente da riciclo. The Italian Institute of Packaging, Milan

21. Kernoghan N (2012) Mineral oil in recycled paper and board packaging. Smithers Pira. https://www.smitherspira.com/testing/food-contact/news-free-webinar-mineral-oil-in-recycled-paper-and-board-packaging.aspx. Accessed 11 Oct 2013
22. European Chemicals Agency (2006) Authorisation list. http://echa.europa.eu/web/guest/addressing-chemicals-of-concern/authorisation/recommendation-for-inclusion-in-the-authorisation-list/authorisation-list. Accessed 18 Oct 2013
23. European Chemicals Agency (2008) Candidate list of substances of very high concern for authorization. http://echa.europa.eu/it/candidate-list-table. Accessed 18 Oct 2013

Chapter 5
Chemical and Microbiological Aspects of the Interaction Between Food and Food Packages

Izabela Steinka

Abstract The purpose of this paper has been to present the interactions between micro-organisms, food containers and packaged foods. The subject of assessment has been the impact of food packaging-related factors on properties of containers and the behaviour of food micro-organisms. In particular, one of the main questions concerns the role of food packages as a source of micro-organisms with the consequent food contamination. In addition, the influence of technological microflora on the packaging features should be discussed and analysed. In relation to these topics, the importance of microflora adhesion and the formation of biofilms on the inner surface of food packages are critical factors. The damage of packages and the possibility of mathematical modelling of micro-organism permeation dynamics through the leak have also been presented. Moreover, the impact of packaging systems and the chemical typology of food contact approved materials have been presented in the context of the preservation of micro-organisms inside containers.

Keywords Food contact material · Food preservation · Mathematical modelling · Mechanical damage · Microbial ecology · Packaged food

Abbreviations

CFU	Colony-forming unit
Φ	Diameter
EVOH	Ethylene vinyl alcohol
C_H	Inhibition coefficient
L	Length
LDPE	Low-density polyethylene
PA	Polyamide
PE	Polyethylene
PET	Polyethylene terephthalate
PP	Polypropylene
PS	Polystyrene
PVC	Polyvinyl chloride

© The Author(s) 2015
C. Barone et al., *Food Packaging Hygiene*, Chemistry of Foods,
DOI 10.1007/978-3-319-14827-4_5

5.1 Introduction

Packages are an integral part of packaged foods; these necessary 'accessories' are designed to function as a protective barrier for foods in terms of quantity and preservation. However, it has been estimated that packages can represent also

1. A source of food contamination
2. A permanent location for microbial spreading because of the existence of a sort of 'gap' or empty space for micro-organisms.

Moreover, packages may be the layer that promotes the interaction between food contact surfaces and packed foods.

At present, the packaging market appears to be dominated by plastic-made containers and objects. Modern environmental requirements force food packaging manufacturers to modify basic materials with the aim of supplying easily biodegradable packages. On the other hand, this type of packaging can also create good or acceptable conditions for the development of food degrading microflora.

In relation to the evaluation of the impact of packages on foods, an important element is the observation of the microbial behaviour when micro-organisms are in contact with packaging surfaces. The interaction between packages and microflora can influence food products in terms of safety and quality.

Microbes in contact with packaging materials may, after a more or less prolonged contact, inhibit their development. On the other side, there is a possibility of penetration into packaged foods. Microflora can also (a) adhere to both surfaces of the same package and (b) form biofilms. In detail, a remarkable modification in the development stage of micro-organisms in concomitant contact with packages and foods may occur. Subsequently, the microbial spreading can occur in packaged products with the typical metabolism of degrading micro-organisms. Sometimes, the contact of micro-organisms with packages is responsible for similar reactions; the delamination of laminates used for food packaging can occur.

5.2 Packaging Materials As a Source of Microflora in Foods

Packaging materials can be a source of microflora. This involves mostly packages obtained from natural materials. At present, there is a certain lack of information with reference to the microbiological quality of packaging materials when intended for contact with foods.

Recent studies have revealed that paper pulp may be a significant source of different bacteria and fungi with the ability of colonising and modifying the structure of paper materials. Many of these life forms with the ability of decomposing cellulose fibres are bacteria belonging to the *Bacillus subtilis* species. Available data show that about 1 % of micro-organisms found in board materials have been originated by simple transfer [1].

Table 5.1 The number and type of micro-organisms present on the outer layer of recycled paper packages [60]

Type of micro-organism	Microflora on the outer layer of packaging materials, CFU/g
Aerobic bacteria, endospores	10^2–10^3
Anaerobic bacteria, endospores	0
Aerobic bacteria	10^3–10^6
Filamentous fungi	10–10^3
Yeasts	0
Mesophilic actinomycetes	0
Thermophilic actinomycetes	10–10^2

The number of fungi and aerobic bacteria in the surface layer of papers can range from 10^3 to 10^6 cells in a gram (Table 5.1). Endospores of aerobic bacteria in superficial layers are also detectable, but the related population does not exceed 10^3 colony-forming units (CFU)/g. In addition, the normal colonisation by viable microflora in food packaging papers consists of bacteria forming thermostable endospores. With specific relation to these bacteria, the amount of endospores found in food contact papers appears to range from 50 to 100,000 CFU per gram of paper [2].

The use of food packaging materials obtained from paperboard coated with mineral oil may implicate the presence of different *Bacillus* and *Peanibacillus* species inside the packaging material. Recent analyses have shown that 90 % of bacteria isolated from paper and cardboard packaging for food applications are assignable to *Bacillus*, *Paenibacillus* and *Brevibacillus* species [3]. In addition, these studies have also proved that food contamination from paper or paperboard is also a result of the contact with internal surfaces with dust on the raw edge [3].

Other data have revealed a significant presence of the following bacteria in paperboard materials: *B. megaterium*, *B. licheniformis*, *Bacillus cereus*, *P. pumilus*, *P. marcerans* and *P. polymyxa* [4].

The presence of *Bacillus* sp. on the inner surface of skived carton packages is probably caused by skiving processes [3]. Recent observation has revealed that the contamination of milk stored in skived 'chemical thermo-mechanical pulp' cartons was higher if compared with milk stored in non-skived cartons. Similar contamination by *B. cereus* has been observed in 10 % of samples obtained from packaging boards used for beverages [3].

It has also been reported that the number of bacteria found in paper or paperboard for beverages appears to depend mainly on two factors: (a) the technological modification of packaging materials and (b) the number of layers in laminates [5]. In relation to recent studies, the main species of bacteria found in a significant number of cardboards have been identified as *B. polymyxa*, *B. circulans* and *B. macerans*. Moreover, the presence of strains of *B. polymyxa* group in carton has been reported to be more significant for the quality of packed foods [5].

According to these data, *B. brevis* seems frequently isolated from samples of these packaging materials if compared with *B. cereus* [5]. In addition, *B. pumilus* has been often reported with remarkable frequency.

By a general viewpoint, the scientific literature shows that the exposure of paper or cardboard packages to food contact may be the reason for the differing dynamics of movements when speaking of aerobic micro-organisms contained in their mass [4]. The amount and dynamics of the moving population appears to depend significantly on the density of packaging materials and the technological protection against dampness. In addition, certain bacteria have demonstrated different attitudes to the survival in a mass of paper or paperboard, depending on the technological degree (or quality) and chemical modifications of cellulosic supports. It has also been reported that dominant species in paperboard materials coated with mineral oil appear to be *B. megaterium* and *B. licheniformis*. The abundant presence of *B. cereus* and *B. coagulans* has been observed in high-density papers. Probably, mineral pigments—used as coating protection for paperboard—may be an excellent source of nutrients [4] for similar bacteria.

For this reason, at least, coated paperboard may be considered an important source of contaminating microflora in foods, when speaking of packages swelling under the influence of pervasive and percolating aqueous solutions.

The detection of *Bacillus* spp. has been demonstrated in other situations. With reference to laminated constituted by paperboard and polyethylene (PE), the dominant microflora seems to be mainly represented by *B. subtilis* and *B. pumilus* types. Moreover, it has been reported that microbial aggregations (obtained from spore germination with bioavailable water) have been found on borders between different layers. The number of observed micro-organisms has been 100–2,000 times greater in these laminates than in cellulose fibres [4].

The penetration of the aqueous phase into paperboard or paper materials causes the migration of microflora or spore germination with the consequent movement towards packaged foods. The dynamics of the above-mentioned movement depends on the thickness and the modification of packaging materials.

The migration of micro-organisms from packaging material (paper, cardboard) is also dependent on the amount of the aqueous phase in food products. The so-called soaking defect of damaged packages can cause the permeation of micro-organisms when speaking of stored foods at room temperature. An example might be correlated to baby functional foods, also named 'baby foods'. The number of identified microbes can range from 10^2 to 2×10^6 CFU/g depending on the type, density and modification of paper and paperboard [4].

This type of interaction, connected with the microbial behaviour, is the result of two concomitant factors:

1. The ability of micro-organisms to survive in the inner structure of packaging materials, and
2. The composition of packaged foods.

In detail, it has been recently reported that even 80 cells can be isolated from mineral-coated paperboard after 2 days of exposure to foods. Apparently, the dynamics of microbial transition from packages towards foods rises after 7 days [4].

Analysed data have shown that bacilli and cocci may use substances contained in lacquers as nutrients. This behaviour may be the cause of the intense development and could explain the dynamics of their infiltration (movements towards foods) when speaking of PE-coated food packaging board. The predictive study of food exposure on the resident microflora in cardboard laminates confirms that the probability of microbial penetration can vary depending on the specific nature of coating layers.

Souminen and co-workers have reported that the total number of living heterotrophic micro-organisms can reach 1,000 CFU in a carton package of 30 g, designed for packaging one litre of milk. With concern to the possibility of microbial transfer from the carton to milk, a high degree of probability has been declared depending on the nature of the cover layer [4].

The number of fungi isolated from cardboard intended for packing of juices and other beverages may show high numbers of filamentous moulds, ranging from 8×10^4 to 2×10^6 CFU/g. The use of mineral oil as superficial coating for cardboard determines the reduction of moulds up to 1,000 times [6–8]. On the other side, moulds appear lower when speaking of cartons for milk and fruit drinks: related numbers are reported between 14 and more than 10^3 CFU/g, respectively [6–8].

The food exposure to fungi present in cardboards is relevantly dependent on the thickness of packages. A peculiar study has been carried out in relation to the possibility of milk contamination by inoculating *Penicillium spinulosum* in paperboard packaging. This research has shown that the probability of the presence of *P. spinulosum* in milk stored at 7 °C for 10–60 days decreases in relation to the increase of the packaging thickness [9].

Other data are available for gable top paperboard cartons for citrus juices and their role as the source of *Penicillium* and *Aspergillus* species, although the microflora is not mainly represented by these moulds [8]. Researchers have concluded that moulds could be isolated from sterilised paperboard carton materials intended for orange juices [10].

Further observations have concerned the assessment of the impact of different environmental conditions on the mould contamination of juices and drinks from cartons [10]. In detail, it has been observed that laminates composed of paperboard and PE-coated paperboard can also be a source of spores and vegetative forms of anaerobic and relatively anaerobic bacteria [7]. Spore aerobic bacteria represent the dominant microflora, and their number is 10 times higher than the amount of spore anaerobic micro-organisms. Occasionally, several *Enterobacteriace* cells have been identified.

The available literature shows also that the safety of packaged foods in cardboard container depends on the type of transferred aerobic bacilli. The presence of *B. cereus* over 100 CFU per gram of cardboard (containers for milk) can be the cause of food poisoning [4]. Another research has indicated that 26.3 % of isolated life forms from food packaging boards appear to be foodborne pathogens such as *B. cereus* [11, 12].

It has been reported that paper packaging for food with low water amounts can be a source of microflora. Generally, the number of resident bacteria in paper used for packing sugar can reach up to 1.2×10^3 CFU. The same type of packaging material can be a source of filamentous moulds when used for wrapping soft candies and similar products. For example, the observed total count of fungi in boxes for confectionery or paper bags for sugar is less than 1.1×10^1 CFU/g [13]. These studies have also shown that packing solutions for pizza products can be a source of bacteria ranging from 1.1 to 1.8 × 10^2 CFU/g [13]. However, the count of micro-organisms isolated from the area of packaging materials depends on sampling methods also. The total amount of bacteria in paperboard by defibreing method appears to be between 10^5 and 10^6 CFU/g [13].

5.3 Survival Rate of Micro-organisms on the Surface of Various Packaging Materials

The survival rate of micro-organisms on the inner surface of packages can vary depending on the properties of packaging materials and predominant conditions on food contact surfaces.

In detail, the dynamics of survival of micro-organisms seems to depend on the type of micro-organisms and environmental conditions (humidity, temperature) on the external surface of packages. As an example, it has been reported that the survival rate of viruses on the surfaces of plastic packages does not exceed 2 days for polio and hepatitis A viruses. Survival times for rotaviruses on glass have been signalled up to 12 days [14]. In addition, hepatitis A viruses may be present up to 60 days on external surfaces of aluminium cans. The survival of *Ortomyxoviridae* on surfaces of steel containers can reach a period of 48 h [15]. Low-density paper does not constitute a good medium for such viruses as *Ortomyxoviridae* (types A and B).

It has been also reported that cooling temperatures and relative humidity values between 25 and 50 % can support the survival of rotaviruses on papers even up to 10 days. However, the same survival does not exceed 48 h at 22 °C if the relative humidity is 85 % [16–18].

The survival rate of filamentous moulds on various packaging materials for dairy products depends mostly from the composition of bioaerosol in storage rooms. Moulds have been found on the surface of PE/polyamide (PA) foils and polystyrene (PS) trays after 24 h-exposure with amounts between 1.47 and 1.84 CFU/cm^2. In addition, parchment paper and aluminium foils may show 1.51 and 1.04 CFU, respectively, when speaking of mould counts [19].

The survival rate of pathogenic bacteria on the surface of plastic materials indicates significant differences in the viability of micro-organisms. A high degree of survival on PS has been observed for *Staphylococcus aureus* [20]. The number of *S. aureus* living cells on the surface of paper and aluminium materials appears to show a decreasing tendency (about 30 %) during storage.

5.4 Adhesion and Formation of Biofilms on Packaging Surfaces

Adsorption is the first stage of the microbial deposition, also named settling, on a specific packaging material. The dynamics of the microbial settling on outer and inner packaging surfaces depends not only on the number of cells but also on important factors such as properties of cell structures.

The presence of cilia in bacteria increases the possibility of adsorption of micro-organisms on both packaging surfaces. In addition, the fixation of the first cells to package surfaces depends on the hydrophobicity and the roughness of supports [21].

The second stage of the microbial settling on packaging materials is adhesion. The assessment of the degree of adhesion is an essential part when speaking of forecasts about the behaviour of the microflora on plastic packages.

Available reported data suggest that the adhesion of micro-organisms to food packages affects many factors, such as superficial electric charges and pH values [22]. Moreover, the interaction between foods in close contact with packages and inner surfaces affects the degree of microbial adhesion [23]. The number of mesophilic aerobes on low-density polyethylene (LDPE) surfaces can most probably reach a maximum amount of 2.38 CFU/cm^2 [22].

Another example concerns *Listeria monocytogenes*: the size of the population of *L. monocytogenes*—capable of adhesion on glass surfaces—is estimated between 5.5 and more than 6.5 log$_{10}$/cm^2 [21, 24].

Each microbial species is characterised by a certain degree of hydrophobicity of external structures. The degree of hydrophobicity depends on the growth phase and the medium where the cells are placed on [25]. In relation to plastic packaging materials, polycationic polymers are considered able to clearly uphold the adhesion of microbial cells. The augment of the degree of adhesion of *E. coli* CSH57 to PS is observed *inter alia* in presence of chitosan [25]. This situation can have a significant impact on the behaviour of bacteria in packaged foods when containers are realised with chitosan as a biostatic substance.

One of the important factors determining the adhesion of micro-organisms to packaging surfaces is the presence of cell adhesion molecules, including exopolysaccharides, by superficial settling. There are many types of cell adhesion molecules: proteins or carbohydrates. An example may be represented by intracellular polysaccharides: this compound is a glucan constituted of β-1,6-*N*-acetyl-glucosamine, produced by *S. epidermidis*. The adhesion of *S. aureus* is favoured by the presence of teichoic acids contained in cell walls [26].

The persistence of micro-organisms on the outer packaging surface depends on the interaction between adhesive bonds and expulsive forces. Should the strength of the adhesion be greater than the strength of 'shearing-off' forces, the development of biofilms and the survival of micro-organisms on outer packaging surfaces would be possible. Otherwise, the microbial persistence on surfaces should not be observable.

Actually, some researchers tend to claim that there is no correlation between these two types of forces: this hypothesis would prove the independence of occurring interactions between bacteria and surface [27]. On these bases, it can be concluded that the adhesion process is not as dynamic as the 'shearing-off' process of micro-organisms from packaging surfaces, depending significantly on the presence of water molecules. Available studies suggest that forces responsible for adhesion are greater in the initial step than at the time of the effective deposition of micro-organisms on surfaces.

The strength of microbial adhesion to inner packaging surfaces must be assessed with regard to the interaction between them and food ingredients. Generally, forces such as hydrogen bonds and van der Waals interactions affect the interactions between micro-organisms (and food ingredients) and packaging surfaces. For example, the strength of the adhesion of E. coli to PS outer surface is reported to reach 4.7 ± 0.6 nN, while the presence of various compounds on this surface causes the reduction of this value even of more than 50 % [28]. The movement of micro-organisms on packaging surfaces at the air–water interface (bioaerosol) requires the force of at least 10^{-7} N [27]. The presence of air–water phases may be also the reason for different values of adhesive strength values of bacteria to both packaging surfaces. Should the outer surface be studied, relevant interactions between the non-food support and surface structures of microbial cells have to be considered.

The estimation of the adhesion energy per area unit between two flat surfaces may be calculated by means of Eq. 5.1 in accordance with Thio and Meredith [28].

$$\Delta_\gamma = \frac{A_H}{12\pi D_0^2} \tag{5.1}$$

where

W interaction energy or work of adhesion of a sphere near a planar surface
D_0 separation distance between the particle and surface
A_H non-retarded Hamaker constant.

The boundary value below which the adhesion of bacteria to packages will not occur can be determined. This value has been determined for E. coli on PS and amounts to 2.9–6.7 nN [28].

On the opposite hand, the free energy of adhesion for micro-organisms such as Staphylococcus spp. or Pseudomonas spp. and packaging materials such as PP and polyvinyl chloride (PVC) appears different. For example, free energy of adhesion between Staphylococcus spp. and PE is 4.2 mJ/m^2 [28]. However, it should be noted that the presence of water vapour can occur on the inner surface of packages, depending on storage conditions of packaged foods.

The evaluation of adhesive forces of micro-organisms to the inner surface of packages must also take into account hydrophilic interactions. An example of this kind of interaction can be the adhesion of L. monocytogenes to glass surfaces [21].

The tendency of micro-organisms to adhere to both sides of packages is not always observed [24, 28]. Some studies have suggested that there is no correlation between the number of adhering micro-organisms and the structure of the surface. However, other studies have confirmed, at the same time, the correlation between surface roughness and adhesive strength [29]. In this situation, the intense colonisation of the package tends to increase because of reduced transverse forces.

The roughness of the outer surface of LDPE packages for food applications may determine higher dynamics of biofilm formation by E. coli and S. aureus because of the size of these bacteria [22]. For example, the infection of plastic bottles with bacilli endospores and moulds is caused by roughness of structural materials [30]. Other studies have also suggested that packages made of PE, polypropylene (PP) and PVC show slight roughness with a consequent poor support for microbial micro-organisms [30].

The size of the aqueous layer on the outside of packages can have a notable influence on forces that lead to the detachment of the micro-organism from the outer surface of the packaging for their movements (in the bioaerosol phase). The critical factor that can determine the possible 'shearing-off' of micro-organisms from the outer surface of packages is also the size of microbial cells [31].

According to several researches, the structure of packaging surfaces can significantly influence the process of adhesion of micro-organisms. In detail, the volumetric extension of the aqueous phase in foods is responsible for the adhesion process of micro-organisms to the inner surface of plastic packages. The size of aqueous layers can determine the dynamics of detachment of micro-organisms from food surfaces and the movement towards the packaging surface.

The value of shearing-off forces should be taken into account when speaking of packaging materials with hydrophilic properties. In relation to glass, shearing-off strength can vary—depending on the species of micro-organisms—from 14 to 37 nN. Within the same species—for different strains, e.g. *S. epidermidis*—the difference between shearing-off force values from glass surface can amount to even 6 nN. With concern to *Streptococcus thermophilus,* the maximum shearing-off strength needed to move bacteria in the air bubble phase is 17 nN [27].

These considerations also took into account the powers of surface tension that affect the durability of microbial adhesion to packaging surfaces.

As an example, surface tension value for *S. epidermidis* on glass surfaces amounts to 0.13 ± 0 06 pN (average data); in addition, this surface tension should be redoubled at least when speaking of shearing-off [27].

The contact of foods with package can be seen as the reason for the formation of biofilms composed of surface microflora in packaged products. Certainly, microflora can produce biofilms on the inner side of the packaging in various ways. An important cause of observed differences for the same packaging material is correlated to properties of packaged food products [23].

The percentage of live micro-organisms isolated from the inner surface of PA/PE containers during storage of highly acid foods indicates a significant degree of yeast affinity in comparison with bacteria. The number of yeast cells may constitute 56 % of the initial number in food products.

Among the bacteria in the surface microflora, *Enterococcus* spp. show the ability to produce biofilms in acid food environments. With concern to *S. aureus,* only 18.1 % of the initial number in the packaged product can be found on the inner surface of the container [32].

With concern to biofilm modifications, two significant elements are the storage time and the way of food movements inside the package during transports. For example, the increase of affinity of filamentous moulds to the inner surface of packages has been observed after 7 days of storage for highly acid foods in cooling conditions [23, 32].

In addition, it can be signalled that the formation of biofilms by *Salmonella, Listeria* and *Staphylococcus* species on packaging surfaces requires at least 10^6 CFU [33]. In can be also considered that the biofilm formation is preceded by the increase in the degree of adhesion when foods are in contact with PA/PE surfaces.

For example, *Lactococcus lactis* on the surface of curd cheeses is reported to show a significant increase in the degree of adhesion to inner container surfaces after 21 days of storage. On the other hand, the visual appearance of these cells to package surfaces in optimal conditions is not convincing in relation to the possibility of significant coverage by *lactococci* in further phases of the experience [34, 35].

With concern to pathogens such as *S. aureus*, a certain tendency to decrease in the number of cells on PA/PE surfaces has been observed: 1.5 log CFU/cm^2 after 21 days of contact, temperature: 4 ± 2 °C. Apparently, superficial images of PA/PE packaging (Fig. 5.1) argue the significant adhesion of *bacilli* and the tendency to form biofilms by *Pseudomonas putida* in these conditions [36].

Moreover, the technological microflora such as *Lactococcus lactis* reveals the ability to synthesise proteins that can facilitate the formation of biofilms [37].

The affinity of micro-organisms for specific packaging materials is dependent on the microbial type, the composition and observable conditions for the assessment of this phenomenon. Model studies have shown much higher degree of *L. lactis* adhesion to biodegradable packages such as polylactide and polylactide film covered with silicon oxides in comparison with traditional packages [38]. *Candida* yeasts show a significant increase of adhesion to hermetically sealed PA/PE packages during long storage times for highly acid food. Short-term tests show high degree of adhesion of *Candida glabrata* to biodegradable packages in model conditions if compared with this phenomenon for PA/PE. The behaviour of observed micro-organisms in model conditions on PA/PE surfaces of laminates is described in Table 5.2.

The structure of biofilms on the inner surface of the packaged foods can result from reciprocal interactions between different species of surface microflora [23]. Such dependences have been observed during the creation of multi-species biofilms in a mixture of pathogenic and saprophytic bacteria [39,40]. It seems that biofilms formed by bacteria on the inner surface of packaged foods reveal a

Fig. 5.1 Microbial contamination of food packaging surfaces. *Pseudomonas putida* on PA/PE [36]

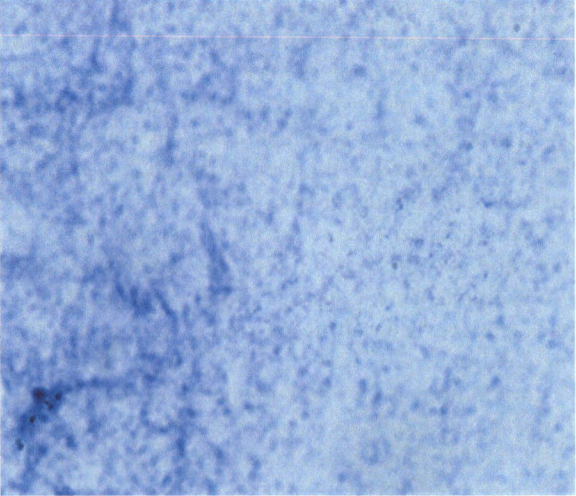

Table 5.2 Adhesion and biofilm on the surface of PA/PE laminates

Type of micro-organism	The effect of superficial contamination (microbial inoculum) on plastic (PA/PE) laminates	
	Inoculum: 10^3 CFU/g	Inoculum: 10^6–10^9 CFU/g
Enterococcus faecalis	No biofilm	No biofilm
Proteus vulgaris	No growth	No growth
Staphylococcus aureus	No adhesion	Slight adhesion of few cells
Candida glabrata	Observed adhesion	Observed biofilm
Candida tropicalis	No cells on the surface	Incipient formation of biofilm
Lactococcus lactis	No adhesion	Several cells are observed on the surfaces

The microbial inoculum influences the final contamination and the formation of related biofilms [36]

diffused nature due to the movement of foods allowing 'falling-off' of cells from the biofilm mass during transport.

The result of this process is the conversion of residual stresses which are felt by aggregates of cells or the flow of condensate of water vapour. Moreover, the lower hydrophobicity of young cells shall be explained by disproportionate changes in the degree of adhesion of bacteria at the time of storage of highly acid foods [23, 34, 35].

5.5 Influence of Packaging Damages on the Behaviour of Micro-organisms

Packaging damages are important factors when speaking of food contamination. Several types of damage responsible for penetration of micro-organisms from the external environment to the food should be distinguished. These are given as follows:

- Mechanical damages
- Perforation
- Delamination
- Flatulence
- Gable top diverging
- Tightness crawling.

Microbial behaviours—in terms of caused damage to packaging—are dependent on various factors. The most important of these points among them are the properties of involved micro-organisms and the type and structure of packaging materials. Important factors that contribute to the formation of damages may be related to too thin layers of laminates, the inappropriate geometry of packages, external pressures and gases produced by micro-organisms.

Depending on whether the perforations have regular or irregular shapes, this penetration of micro-organisms to the inside of the packaging is made possible by different sizes of emerging gaps.

Fig. 5.2 Perforations in a PS
container. Observed damages
appear to exhibit irregular
shapes

In relation to irregular damages (Fig. 5.2), the movement of *E. coli* (initial inoculum: 10^8 CFU) requires a leak with diameters from 22 to 175 μm. On the other
hand, the microbial movement can also happen through holes of 'regular' shape
with diameters between 17 and 81 μm (Fig. 5.3).

The transfer of micro-organisms from the outer side of packages to packaged
foods also depends on the pressure of food on the slit, time and temperature of
storage and the nature of the medium inside the crack.

The minimum diameter for penetration of microbes inside hermetically packaged foods can be determined. Equation 5.2, which describes also the difference of
pressures prevailing on both sides of the package, can be examined.

$$D_{\mathrm{H}} = 4\sigma \left(\frac{P_0}{0.39} + \rho g L \right)^{-1} \tag{5.2}$$

Fig. 5.3 Perforations in a
PS container. Observed holes
appear to exhibit roughly
regular shapes

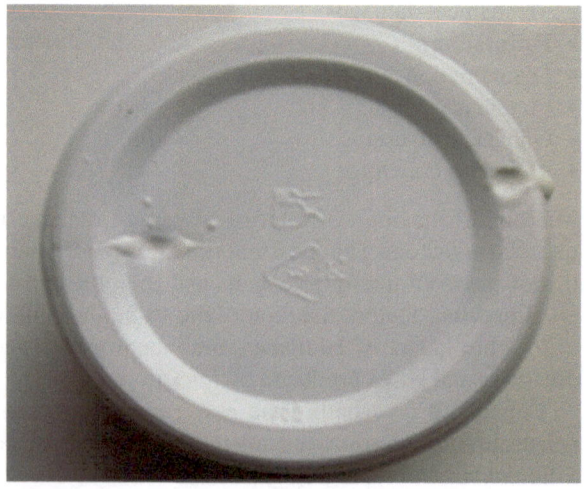

Fig. 5.4 Aqueous condensation inside vacuum packages. The presence of water vapour condensation in closed packages can act as a carrier of microbial contamination

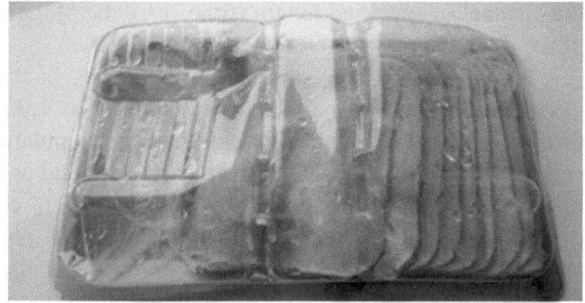

Equation 5.2 is based also on the following terms [40, 41]:

D_H diameter of penetration
σ fluid density should be taken into account
P_0 pressure inside the package
ρ surface tension
g   gravitational acceleration
L   length of the slit.

Food contamination is proportional to the difference of prevailing pressures between the inner surface and the outer environment. The further movement of micro-organisms is dependent on the size and the amount of condensate on the inner surface of the container (Fig. 5.4).

The dynamics of displacement of bacteria through the package in liquid media depends not only on the size of the diameter of the leaking channel, but also on the shape of the channel. Morphology and abundance of bacteria are other factors that significantly influence the degree of contamination of food when slits occur in packages.

With reference to *Enterobacter cloacae* and *Enterobacter aerogenes*, the dynamics of microbial penetration, expressed as log cells/channel/s, is greater for a channel of size of 0.78–120 μm^2 if the pressure amounts to 51–305 mm Hg [42].

However, other studies [43] have shown, among other things, that properties of micro-organisms are responsible in a greater extent than pressure values with concern to the microbial penetration through leak channels in packaging materials.

The presence of numerous cilia allows micro-organisms the easier penetration through leak channels in packaging materials, if compared to motionless bacteria. An example can be *Pseudomonas fragi*: its dynamics of penetration does not vary for different pressures and sizes of leaks.

At the same time, recent studies have shown that the penetration of various micro-organisms such as *Pseudomonas, Staphylococcus* and *Bacillus* spp. on liquid medium is similar through the slit of little dimensions [42].

An important effect in the process of microbial penetration to the inside of the container is related to the number of bacteria. It has been proved that the population of motile bacteria such as *Pseudomonas* spp. with fewer cells penetrates more easily than the one that contains more than 10^4 CFU of cells [42].

This dependence is also observed when speaking of aerobic bacteria such as *Bacillus* species. A population exceeding the number of 4×10^9 cells can cause, in the channel of specified length (time: 1 s), the plugging of the leak and a rapid fall of the dynamics of microbial penetration [42].

This phenomenon is observed for various packaging materials. It has been clearly suggested that populations consisting of 1,000 cells penetrate more easily through even twice shorter cracks than the population consisting of one million cells [44]. This includes, *inter alia*, bags designed as tri-ingredient laminates composed of PP/PA/PP layers with slits of 3- and 6 μm-cells [44].

This effect, associated with the dynamics of population transfer through the leaks in packages, is also dependent on the kind of existing micro-organisms and their mutual proportions [42]. In relation to the same number of micro-organisms in mixed populations, it has been reported that the microbial number of *Staphylococcus* bacteria can pass (with condensate) through the leak channel with augments of 4–6 times in comparison with the number of *Pseudomonas* and *Bacillus* species.

Contrary to the opinions of some researchers, it seems that the composition—in terms of types in a mixture of bacteria population—exhibits more significant impact on the dynamics of penetration than the ability to produce exopolysaccharides.

5.6 Determinants of Microbial Penetration on a Liquid and with Aerosol Medium

The probabilistic prediction of the movement of micro-organisms to the inside of the container is based on the assumption about the pressure difference: this difference can be calculated by means of Eq. 5.3:

$$\Delta P = P_0 - P_L + \rho g L \tag{5.3}$$

where

P_0 pressure inside the package
P_L external pressure
G gravitational acceleration
P surface tension
L length of the slit.

On the other hand, the following terms should be taken into account (Eq. 5.4):

r radius of drops (pushed out by the pressure)
σ fluid density

$$\Delta P = \frac{2\sigma}{r}, \tag{5.4}$$

Therefore, the knowledge of atmospheric pressure (P_{atm}) allows taking into account all the factors that determine the transfer of micro-organisms on drops of fluid to the inside of the packaging (Eq. 5.5):

$$P_L = P_{atm} + \frac{2\sigma}{r} - \rho g L, \tag{5.5}$$

where P_{atm} denotes the atmospheric pressure.

The assessment of the degree of contamination of foods by moving molecules can be expressed by means of Eq. 5.6:

$$n = \frac{N\left(\frac{R}{Z+1}\right)^2}{8\mu[L(A+1)]^2} \cdot \left(P_0 - P_{atm} - \frac{2\sigma}{r} + \rho g L\right) \tag{5.6}$$

where

n total number of microorganisms
N initial number of microorganisms
Z plasticity of packaging materials
A rigidity of the packaging material
R radius of fluid drops
μ dynamic viscosity
L length of the slit.

However, in order to depict the ability of microbial movement through a barrier which is the package to the inside of it, the hydrophobicity of packaging materials should be, as well, carefully considered.

The final equation showing the movement probability for a certain number of micro-organisms through the leak in the package takes the form of Eq. 5.7:

$$n = \frac{N\left(\frac{R}{Z+1}\right)^2}{8\mu[L(A+1)]^2} \cdot \left(P_0 - P_{atm} - \frac{2\sigma}{r} + \rho g L\right) \cdot (1 + \cos\theta) \cdot \frac{D_A}{D} \tag{5.7}$$

where

θ angle of moistening required to assess the hydrophobicity of the packaging material
D molecular radius
D_A distance of molecules from the surface of the material.

The penetration of micro-organisms to the inside of the packaging can happen with aerosol in the absence of liquid media. At the same time, the microbial movement to the inside of packages with aerosol requires taking into account other factors. These parameters are, *inter alia*, the density of aerosols, the size of aerosolised molecules and micro-organisms, the environmental moisture and the amount of microbial populations.

Studies by some authors have shown varied results. For example, a certain microbial penetration has been reported and statistically demonstrated [45] through leaks of radius between 10 and 20 μm and length of 5–10 mm for *P. fragi* if present in aerosols (radius: 2.7 μm).

Other studies have suggested that higher difference in pressures causes slight reduction in the number of penetrating micro-organisms with the same size of the crack [46].

The increase of crack radius causes the growth of the dynamics of penetration together with the augment in the difference of pressures between interior and exterior walls of the packaging. In relation to cracks with radius between 2 and 50 μm, the number of moving bacteria is 10^4 CFU when vacuum pressure differentials range between -34.5 and -6.9 kPa [46].

Different researchers have also suggested that the dynamics of the infiltration through the cracks in packages falls together with time [47]. Equation 5.8 allows the prediction of the movement of micro-organisms in aerosol through leak in packages.

$$n = \frac{N\left(\frac{R}{Z+1}\right)^2}{8\mu[L(A+1)]^2} \cdot \left(P_0 - P_{\text{atm}} - \frac{2\sigma}{r} + \rho g L\right) \cdot C_H \cdot \frac{D_A}{D} \tag{5.8}$$

where C_H denotes the inhibition coefficient.

Equation 5.8 takes into account the inhibition coefficient (C_H) and the distance of molecules from packaging surfaces, D_A. The remaining parameters of Eq. 5.8 have been described with reference to Eq. 5.7.

An additional parameter, which should be taken into account, is the possibility of aggregation of the molecules over the leak. Equation 5.9 has been applied to describe the aggregation of molecules. The potential energy of interaction is given by

$$E_P = \frac{-a}{d^{m-1}} + \frac{b}{d^{n-1}}, \tag{5.9}$$

wherein $\frac{-a}{d^{m-1}}$ and $\frac{b}{d^{n-1}}$ are two components corresponding to the magnetic force $\frac{-A}{d^m}$ and the repulsive force $\frac{B}{d^n}$, respectively. d represents the distance between molecules. A, B and a, b couples are constants dependent on the kind of interacting molecules, and m is a power that best describes the dependence of potential on the distance d. Should the long-range interaction result only from the interaction of van der Waals bonding, m would be equal to 6. The power coefficient n of the repulsion force is generally much larger than 6. The balance of forces occurs when

$$\frac{-A}{d^m} + \frac{B}{d^n} = 0 \tag{5.10}$$

which is equivalent to the Eq. 5.11:

$$\frac{B}{A} \cdot d^{m-n} = 1 \tag{5.11}$$

Should Eq. 5.11 give a result higher than one, then repulsive forces would be stronger: this condition is associated with better flow. On the other hand, should Eq. 5.11 give a result lower than one, then magnetic forces would be stronger (poor flow of molecules).

Hence, after taking into account the aggregation of molecules in Eq. 5.8, it takes the form:

$$n = \frac{N\left(\frac{R}{Z+1}\right)^2}{8\mu[L(A+1)]^2} \cdot \left(P_0 - P_{\text{atm}} - \frac{2\sigma}{r} + \rho g L\right) \cdot C_H \cdot \frac{D_A}{D} \cdot \frac{B}{A} \cdot d^{m-n} \tag{5.12}$$

The last parameter, required for describing the movement of bacteria with aerosol to the inside of the packaging, is the difference in temperatures inside and outside the packaging. Taking into consideration the ratio of outside to the inside the package temperature, with an accuracy of a constant k (the effect is proportional to the flow rate of microbes), the final result is obtained. Then, the prediction of the number of aerosolised moving microbes from the external environment into packaged foods can be carried out using Eq. 5.13:

$$n = \frac{N\left(\frac{R}{Z+1}\right)^2}{8\mu[L(A+1)]^2} \cdot \left(P_0 - P_{atm} - \frac{2\sigma}{r} + \rho g L\right) \cdot C_H \cdot \frac{D_A}{D} \cdot \frac{B}{A} \cdot d^{m-n} \cdot k\frac{T_L}{T_0}$$

$$(5.13)$$

where

T_L temperature outside the package
T_0 temperature inside the package.

5.7 Determination of the Minimum Leak for Penetration of Micro-organisms Through the Package

The designation of minimum crack radius values for the promotion of microbial penetration, depending on the nature and the shape of packaging materials, is shown in Table 5.3.

The inflexibility of packages should be mentioned when determining the minimum size of leaks. The minimum radius is much greater in flexible packages than in inflexible or semi-flexible materials. In flexible laminates made of plastics, this radius is 22 μm, while the same parameter is only 5 μm in inflexible boxes. The penetration through inflexible bottles requires a leak ranging from 5 to 15 μm. Similar values are correlated to semi-flexible plastic containers: minimum leaks promoting the microbial penetration range from 5 to 10 μm [48].

Table 5.3 The influence of the dimension of holes in packages on the penetration of micro-organisms into packaged foods [41, 42, 48, 51, 56–59]

Packaging materials	Packaging typology	Size of observed leaks, diameters (Φ) or length (L)
PE, PP	Bags	$\Phi = 20\,\mu m$ or $L = 5\,mm$
Metal (steel, aluminium)	Cans	$\Phi \geq 1\,\mu m$
Steel (steel)	Cans (inner and external pressure values are different)	$\Phi = 2\,\mu m$; $\Phi > 5\,\mu m$
PE/PET/EVOH/PP	Trays	$L = 70–200\,\mu m$
PET	Bottle	$\Phi < 5\,\mu m$
PS	Cups	$\Phi = 10–20\,\mu m$

The permeation of microbes through packages is also significantly dependent on the type of micro-organisms and their classification (Procaryota or Eucaryota). Available researches have shown that ten times more bacteria passes through the leak in comparison with the number of filamentous moulds. For example, should the following bacteria be inoculated, *E. coli, S. aureus, B. spizizenii,* 10^6 CFU/mL, and fungi, 10^5 CFU/mL, the presence of *Candida albicans* and *Aspergillus brasiliensis* inside the package would be stated after 14 days for 15-mm cracks and after 5 days for 20- and 50-µm cracks [49].

The infiltration of bacteria into the interior of drinks through polyethylene terephthalate (PET) packages depends on the density of the fluid. The contamination of food with micro-organisms infiltrating through the leaks depends on both the size of the crack and the consistency and the type of food. Studies suggest that the infection of chicken dishes shows the presence of 10^6 to 10^7 CFU/g, while beef *enchiladas* infection varies from 1 to 3 log CFU/g on condition that holes in PET/ethylene vinyl alcohol (EVOH)/PP trays have the same size [50].

The dynamics of microbial penetration to the inner side of semi-flexible containers depends on the pressure of foods on leakage channels and the tenacity in these channels. The type of packaged foods is also important.

The penetration of bacteria into inflexible PP-made containers depends on the radius of these micro-organisms [51]. Table 5.4 shows the probability of the penetration of various bacteria to the inside of PP containers.

Another important factor that influences the penetration of micro-organisms into closed containers with lids is the surface tension. An important role is also played by the tightness of the leakage channel itself. Bacterial motility is a factor that determines the dynamic of passage of bacteria into the packages [43]. The presence of exopolysaccharides in bacteria causes changes in the tenacity of leakage channels and can reduce the dynamics of penetration into the interior packaging.

For example, in relation to PS container with lids made of aluminium foils (Fig. 5.5), the dynamics of penetration may amount to 1 log CFU/24 h for *E. aerogenes* [51, 52].

The motility of micro-organisms is essential for the penetration of bacteria into inflexible and semi-flexible packages; on the other hand, this factor does not appear so important when speaking of metal containers.

Metal packages show the highest degree of easy microbial penetration inside packages when speaking of minimal leaks (the microbial movement is 2 µm).

Table 5.4 Probability of microbial penetration through PP materials depending on the radius of holes and the micro-organism [51]. Modified data

Type of micro-organism	Microbial contamination (CFU/g) when hole sizes are 10 µm	Microbial contamination (CFU/g) when hole sizes are 20 µm
Bacillus subtilis	5	11
Leuconostoc mesenteroides	0	10
Micrococcus varians	45	70

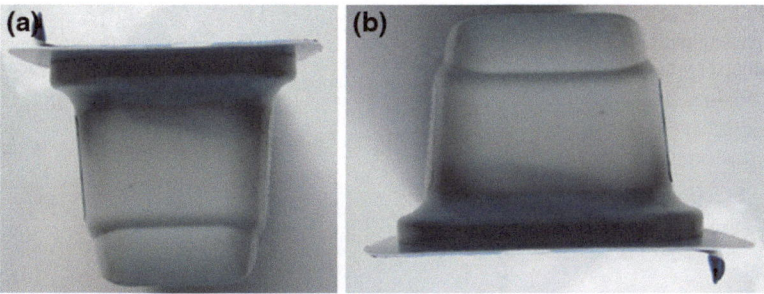

Fig. 5.5 Microbial penetration in food packages. The imperfect closure and the possible crawling between the container and the lid can favour microbial spreading

Recent experiments have shown that the leaking of cells of *P. fragi* (initial inoculum: 10^6 g/cm^3, pressure values between 6 and 20.7 kPa) can happen even when 5-μm-big cracks occur [41].

Glass containers are considered to be high-barrier materials for both biological and mechanical impurities. On the other hand, cracks in this packaging material can allow microbes to move into the package. The presence of 15-μm microleaks may determine the transition of bacterial populations with a speed of 1.3×10^{-5} mbarl/s [47].

5.8   Interactions of Micro-organisms with Packages

Interactions between packages and food can cause diverse modifications on the surface of foods and containers. This interaction concerns mainly wrapped, non-hermetically packed foods or on trays with different types of materials (Fig. 5.6).

Sometimes, a double packaging system is the cause of the growth of microflora responsible for qualitative changes of packaged foods. Figure 5.7 shows a peculiar white cheese packed in parchment paper and also hermetically sealed in PA/PE films.

Fig. 5.6 Biofilm formation by *Aspergills flavus* on cellophane surfaces after contact with cheese

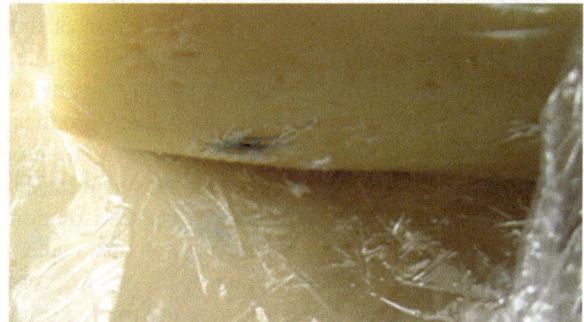

Fig. 5.7 Flatulence in
PE/PA packaging. Gases
are produced by *Candida
guilerimondii*

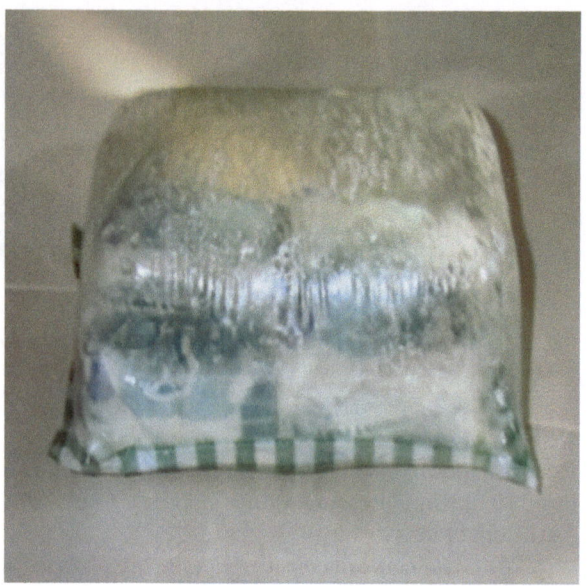

Fig. 5.8 Sliced meats
packaged into polystyrene
boxes. *Brochothrix
thermosphacta* spoilage and
consequent gas production

Products of the metabolism of surface microflora are responsible for external flatulence of the pack [23]. This is the effect of packaging barriers against the outer layer protecting from the leakage of the aqueous phase from the product.

In addition, the use of boxes for packing products of animal origin without gas scavengers systems (sachets) can be the cause of flatulence of packages (Fig. 5.8), probably as a result of the activity of proteolytic microflora [53].

The wrapping of foods on PS trays may favour the growth of aerobic microflora (Fig. 5.9). Compared to hermetically packed products, loss of freshness in foods on PS trays is essentially due to the presence of slime on surfaces, smell changes, and higher number of filamentous moulds on surfaces [23]. The wrapping of products of plant origin in foils, placed on trays, causes isolation from products of filamentous moulds as well as enterococci, staphylococci and *E. coli*, lowering food safety expectations. This situation is the result of the presence of significant amounts of oxygen in the space between the tray and the foil used to wrap food [54].

Fig. 5.9 *Blue cheese* on PS tray wrapped with plastic foil

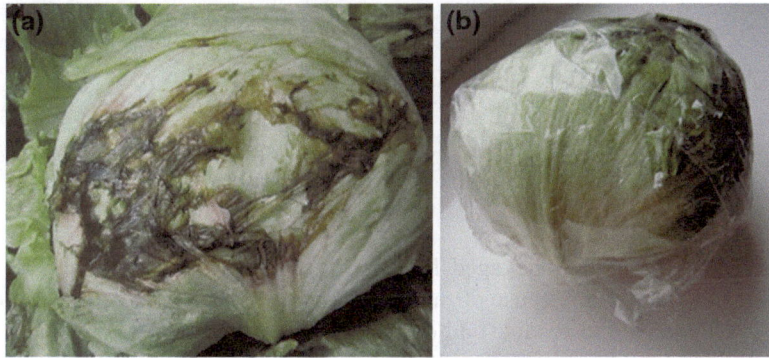

Fig. 5.10 The impact of different foils for wrapped vegetables. Effect of high-barrier foils on the development of *Pseudomonas cichorii* (**a**). Effect of low barrier-perforated foils for wrapped iceberg lettuce (**b**)

The wrapping of products of animal origin with a double layer of paper and parchment does not protect from water leakage; in addition, the germination of *Bacillus* spp. spores from packaging materials may be stimulated during 48-h storage periods.

Only the appropriate selection of perforated PE packaging materials used for iceberg lettuce packaging may prevent further development of psychotropic agents responsible for the formation of putrefactive changes of products (Fig. 5.10). It should be also noted that the choice of hermetic food packaging may cause various microbial changes.

As an example, in relation to vacuum-packaged products, vacuum parameters are responsible for the possible microbial inhibition. Several data have shown that the strict adhesion of PE meat surfaces may determine the development of lactic fermentation bacteria such as *Lactobacillus* spp. (Fig. 5.11). As a result, many products such as diacetyl or acetone can be obtained with consequent biochemical changes [55].

Fig. 5.11 Hermetic
packaging for beef products.
High-vacuum packages may
efficacy contrast undesirable
biochemical changes

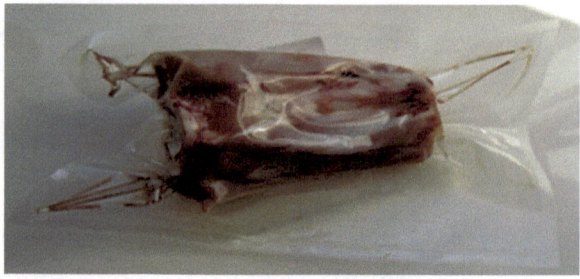

Fig. 5.12 Niche created in a
hermetic container between
food and packaging surfaces

In relation to vacuum-packaged foods with high water content such as lactic acid cheeses, the possible cause of microbiological changes may be the size of the empty space (above the product). In detail, the volume of the aqueous phase flowing from curds could not be carried away due to the consistency of these products. As a result, both aerobic micro-organisms and microphilic bacteria micro-organisms may be favoured in these conditions because of a residual air bioavailability (Fig. 5.12).

A remarkable number of biosynthesised products into hermetically packaged foods are responsible for deformations of packages. This phenomenon can be particularly observed in fermented milk drinks packaged in containers with welded lids.

5.9 Forecasting the Stability of Packaging Materials

The suitability of packaging materials for safe food packaging may be predicted. Packaging materials intended for food contact applications must meet many requirements in accordance with international quality standards and regulatory norms. The basic premise is the warranty of a certain 'barrier effect' not only with concern to physical factors, but also with reference to the adequate 'sterility' of packaged foods.

The promotion of adhesion of micro-organisms and the increased tendency to form biofilms is a factor disqualifying packaging materials. In relation to these problems, there is the possibility of using mathematical models to assess the suitability of packaging materials in the context of its interaction with foods and micro-organisms.

An example can be proposed for the evaluation of the microbiological stability of PA/PE foils intended for low-acid food packaging at pH 4.5 [38]. Equation 5.9 can be used for this purpose:

$$S(t) = 0.2D(t) + 0.3P(t) + 0.4B(t) + 0.1T(t) \tag{5.14}$$

where

D	degradation of the material (described by film absorption and migration)
T	shelf life of packaging
B	biostatic properties
T	time
a, b, c, d	coefficients of pertinence parameters.

This model has been proposed as the basis for the assessment of the suitability of packaging materials of any kind intended for food contact applications.

References

1. Raaska L, Sillanpää J, Sjöberg AM, Suihko ML (2002) Potential microbiological hazards in the production of refined paper products for food applications. J Ind Microbiol Biotechnol 28(4):225–231. doi:10.1038/sj/jim/7000238
2. Ekman J (2011) Bacteria colonizing paper machines. Dissertation, University of Helsinki
3. Pirttijarvi T (2000) Contaminant aerobic sporeforming bacteria in the manufacturing processes of food packaging board and food. Dissertation, University of Helsinki
4. Souminen I, Suihko ML, Salkinoja-Salonen M (1997) Microscopic study of migration of microbes in food-packaging paper and board. J Ind Microbiol Biotechnol 19:104–113. doi:10.1038/sj.jim.2900424
5. Valsanen OM, Mentu J, Salklnoja-Salonen MS (1991) Bacteria in food packaging paper and board. J Appl Bacteriol 71:130133. doi:10.1111/j.1365-2672.1991.tb02967.x
6. Suihko ML, Skytta E (1997) A study of the microflora of some recycled fibre pulps, boards and kitchen rolls. J Appl Microbiol 83:199–207. doi:10.1046/j.1365-2672.1997.00219.x
7. Kneifel W, Kaser A (1994) Microbiological quality parameters of packaging materials used in the dairy industry. Arch Lebensmittelhyg 45:25–48
8. Narciso JA, Parish ME (1997) Endogenous mycoflora of gable-top carton paperboard used for packaging fruit juice. J Food Sci 62(6):1223–1239. doi:10.1111/j.1365-2621.1997.tb12249
9. Sammons LD (1999) Migration of *Penicillium spinulosum* from paperboard packaging to extended shelf life milk. Dissertation, Virginia Polytechnic Institute and State University
10. Narciso JA, Parish ME (2000) Relationship of mold in paperboard packaging to food spoilage. Dairy Food Environ Sanit 20(12):944–951
11. Suihko ML, Stackebrandt E (2003) Identification of aerobic mesophilic bacilli isolated from board and paper products containing recycled fibers. J Appl Microbiol 94(1):25–34. doi:10.1046/j.1365-2672.2003.01803.x
12. Priha O, Hallamaa K, Saarela M, Raaska L (2004) Detection of *Bacillus cereus* group bacteria from cardboard and paper with real-time PCR. J Ind Microbiol Biotechnol 31(4):161–169. doi:10.1007/s10295-004-0125-x

13. Guzinska K, Owczarek M, Dymel M (2012) Investigation in the microbiological purity of paper and board packaging intended for contact with food. Fibres Text East Eur 20(6B96):186–190. Available http://fibtex.lodz.pl/2012/6B/186.pdf. Accessed 28 Oct 2014
14. Rzeżutka A, Cook N (2004) Survival of human enteric viruses in the environment and food. FEMS Microbiol Rev 28:441–453. doi:10.1016/j.femsre.2004.02.001
15. Noyce JO, Michels H, Keevil CW (2007) Inactivation of influenza A virus on copper versus stainless steel surfaces. Appl Environ Microbiol 73(8):2748–2750. doi:10.1128/AEM.01139-06
16. Bean B, Moore BM, Sterner B, Peterson LR, Gerding DN, Balfur HH (1982) Survival of influenza Viruses on environmental surfaces. J Infect Dis 146(1):47–51. doi:10.1093/infdis/146.1.47
17. Sattar SA, Lloyd-Evans N, Springthorpe VS, Nair RC (1986) Institutional outbreaks of rotavirus diarrhoea: potential role of fomites and environmental surfaces as vehicles for virus transmission. J Hyg (Lond) 96(2):277–289. doi:10.1017/S0022172400066055
18. Kramer A, Schwebke I, Kampf G (2006) How long do nosocomial pathogens persist on inanimate surfaces? A systematic review. BCM Inf Dis 6:130. doi:10.1186/1471-2334-6-130
19. Steinka I, Przybyłowski P (1998) Jakość mikrobiologiczna kwasowych serów twarogowych a metody pakowania. Przem Spoż 11:47–49
20. Tiller JC, Liao CJ, Lewis K, Klibanov AM (2001) Designing surfaces that kill bacteria on contact. PNAS 98(11):5981–5985. doi:10.1073/pnas11143098
21. Teixeira P, Silva S, Araujo F, Azeredo J, Oliveira R (2007) Bacterial adhesion to food contacting surfaces. In: Méndez-Vilas A (ed) Communicating current research and educational topics and trends in applied microbiology. Microbiology series 1 vol 1. Formatex, Badajoz, pp 13–20. Available http://www.formatex.org/microbio/pdf/Pages13-20.pdf. Accessed 27 oct 2014
22. Silva CAS, Andrade NJ, Soares NFF, Fereira SO (2003) Evaluation of ultraviolet radiation to control microorganisms adhering to low-density polyethylene films. Braz J Microbiol 34(2):175–178. doi:10.1590/S1517-83822003000200017
23. Steinka I (2003) Wpływ interakcji opakowanie—produkt na jakość mikrobiologiczną hermetycznie pakowanych serów twarogowych. Wydawnictwo Akademii Morskiej, Gdynia
24. Silva S, Texeira P, Oliveira R, Azeredo J (2008) Adhesion to and viability of Listeria monocytogenes on food contact surfaces. J Food Protect 71(7):1379–1385
25. Goldberg S, Doyle RJ, Rosenberg M (1990) Mechanism of enhancement of microbial cell hydrophobicity by cationic polymers. J Bacteriol 172(10):5650–5654
26. Naber CK (2009) Staphylococcus aureus bacteremia: epidemiology, pathophisiology, and management strategies. Clin Inf Dis 48(Suppl 4):S 231–237. doi:10.1086/598189
27. Boks NP, Norde W, van der Mei HC, Busscher J (2008) Forces involved in bacterial adhesion to hydrophilic and hydrophobic surfaces. Microbiol 154(3):3122–3133. doi:10.1099/mic.0.2008/018622-0
28. Thio BJR, Meredith C (2008) Quantification of E. coli adhesion to polyamides and polystyrene with atomic force microscopy. Colloids Surf B Biointerfaces 65:308–312. doi:10.1016/j.colsurfb.2008.05.005
29. Characklis WG (1990) Microbial fouling. In: Characklis WG, Marshall KC (eds) Biofilms. Wiley, New York, pp 523–634
30. Jeje JO, Oladepo KT (2012) A study of sources of microbial contamination of packaged water. Trans J Sci Technol 2(9):63–76. Available http://tjournal.org/tjst_october_2012/6.pdf. Accessed 28 Oct 2014
31. Busscher HJ, van der Mei H (2006) Microbial adhesion in flow displacement systems. Crit Microbiol Rev 19(1):127–141. doi:10.1128/CMR.19.1.127-141.2006
32. Steinka I (2008) Lactic acid cheese safety. Nova Science Publishers Inc, New York
33. Steinka I, Morawska M (2010) Ocena biofilmu formowanego przez wybrane bakterie i grzyby na powierzchni opakowań stosowanych do pakowania twarogów. Unpublished data
34. Steinka I, Kukulowicz A (2004) Assessment of adherence degree of adhesion of the Lactococcus sp. to surface of PA/PE laminates. Jt Proc 17:51–53. WSM Gdynia, Hochschule Bremerhaven
35. Steinka I, Kukułowicz A (2004) Adhesion of Lactococcus bacteria to the surface of traditional and biodegradable packaging laminates. Polish J Nutr Sci 2:151–156

36. Steinka I, Morawska M (2010) Ocena biofilmu formowanego przez wybrane bakterie i grzyby na powierzchni opakowań stosowanych do pakowania twarogów. Unpublished data
37. Luo H, Wan K, Wang HH (2005) High-frequency conjugation system facilitates biofilm formation and pAMβ1 transmission by *Lactococcus lactis*. Appl Environ Microbiol 71(6):2970–2978. doi:10.1128/AEM.71.6.2970-2978.2005
38. Morawska M, Steinka I, Blokus-Roszkowska I (2013) Modelowanie matematyczne w ocenie jakości materiałów opakowaniowych. Zeszyt Naukowe Akademii Morskiej w Gdyni 80:5–12. Available http://zeszyty.am.gdynia.pl/artykul/Modelowanie%20matematyczne%20w%20ocenie%20jakosci%20materialow%20opakowaniowych_201.pdf. Accessed 28 Oct 2014
39. Tolker-Nielsen T, Molin S (2000) Spatial organization of microbial biofilm communities. Microb Ecol 40:75–84. doi:10.1007/s002480000057
40. Keller S, Marcy J, Blakistone B, Hackney C, Carter WH, Lacy G (2003) Effect of microorganism characteristics on leak size critical to predicting package sterility. J Food Prot 66(9):1716–1719
41. Gnanasekharan V, Floros JD (1994) Package integrity evaluation. Criteria for selecting a method. Part I. Pack Technol Eng 3(6):44–48
42. McEldowney S, Fletcher M (1990) The effect of physical and microbiological factors on food container leakage. J Appl Bacteriol 69(2):190–205. doi:10.1111/j.1365-2672.1990.tb01509.x
43. Keller S, Marcy J, Blakistone B, Hackney C, Carter WH, Lacy G (2003) Application of fluid modeling to determine threshold leak size for liquid foods. J Food Prot 66(7):1260–1268
44. Song YS, Hargraves WA (1998) Postprocess contamination of flexible pouches challenged by in situ immersion biotest. J Food Prot 61(12):1644–1648
45. Keller SW (1998) Determination of the leak size critical to package sterility maintenance. Dissertation, Virginia Polytechnic Institute and State University
46. Gibney MJ (2000) Predicting package defects: quantification of critical leak size. Dissertation, Virginia Polytechnic Institute and State University
47. Morrical BD, Goverde M, Grausse J, Gerwig T, Vorgrimler L, Morgen R, Büttiker JP (2007) Leak testing in parenteral packaging: establishment of direct correlation between helium leak rate measurements and microbial ingress for two different leak types. PDA J Pharm Sci Technol 61(4):226–236
48. Ravishanker S, Maks ND, Teo AYL, Strassheim HE, Pascall MA (2005) Minimum leak size determination, under laboratory and commercial conditions, for bacterial entry into polymer trays used for shelf-stable food packaging. J Food Prot 68(11):2376–2382
49. Pethe V, Dove M, Terentiev A (2011) Integrity testing of flexible containers. BioPharm Int 24(11):42–49. Available http://www.biopharminternational.com/biopharm/article/articleDetail.jsp?id=747047&sk=&date=&pageID=6. Accessed 28 oct 2014
50. Ravishanker S, Maks ND, Teo AYL, Strassheim HE, Pascall MA (2005) Minimum leak size determination, under laboratory and commercial conditions, for bacterial entry into polymer trays used for shelf-stable food packaging. J Food Prot 68(11):2376–2382
51. Hurme EU, Wirtanen G, Axelson-Larsson L, Pachero NAM, Ahvenainen R (1997) Penetration of bacteria through microholes in semirigid aseptic and retort packages. J Food Prot 60(5):520–524
52. Avhenainen R, Mattila-Sandholm T, Axelson L, Wirtanen G (2006) The effect of microhole size and foodstuff on the microbial integrity of aseptic plastic cups. Packaging Technol Sci 5(2):101–107. doi:10.1002/pts.2770050209
53. Steinka I (2012) Opakowania hermetyczne bezpieczeństwo i akceptacja konsumencka. Informator Masarski 6/2012, Masterpress Poradnik, Białystok
54. Steinka I (2009) Assessment of interactions occurring between microflora and packaging applied for food. In: Bellinghouse VC (ed) Food processing: methods, techniques and trends. Nova Science Publishers Inc, New York
55. Nychas GJE, Skandamis PN, Tassou CC, Koutsoumanis KP (2008) Meat spoilage during distribution. Meat Sci 78(1–2):77–89. doi:10.1016/j.meatsci.2007.06.020
56. Gilchrist JE, Rhea US, Dickerson RW, Campbell JE (1985) Helium leak test for micron-sized holes in canned foods. J Food Prot 48(10):856–860

57. Jarrosson BP (1992) Closure integrity of heat sealed aseptic packaging using scanning acoustic microscopy. Dissertation, Virginia Polytechnic Institute and State University
58. Lake DE, Graves RR, Lesnewski RS, Anderson JE (1985) Postprocessing spoilage of low-acid canned food by mesophilic anaerobic sporeforms. J Food Prot 48(3):221–226
59. Sivaramakrishna V, Mehta A, Schramm G, Pascall MA (2007) Leak detection in polyethylene terephthalate bottles filled with water and pulped and unpulped orange juice using a vacuum system. J Food Prot 70(10):2365–2372
60. Steinka I (2011) Mikrobiologia żywności i artykułów przemysłowych. Wydawnictwo Akademii Morskiej, Gdynia

Chapter 6
Basic Principles of Corrosion
of Food Metal Packaging

Angela Montanari

Abstract The corrosion of metal packs is of major importance for health reasons and with reference to the possible reduction of shelf-life values. Basically, main failures of metal packages can be excessive metal amounts in food products, hydrogen swelling, perforation, lacquer blistering or delaminating, and modification of sensorial properties. Therefore, the possibility of minimising corrosion phenomena is of great concern depending on the exact knowledge of chemical and physical factors and causes. This chapter examines the thermodynamic and kinetic aspects of the corrosion mechanisms of tinplate, tin-free steel (TFS) and aluminium with a brief introduction to corrosion theory. In detail, a description of main anodic, cathodic and galvanic coupling prevailing reactions is provided in this chapter with particular reference to preserved foods and possible consequences (aggressiveness). The following factors with some correlation with corrosive phenomena are considered: chemistry of the metallic material, food formulation, packaging process, properties of the organic coating, and shape and capacity of the container. In particular, the role of oxygen is discussed. In addition, the description of the corrosion morphology is shown along with some practical examples with reference to failures such as detinning and pitting.

Keywords Aluminium alloy · Detinning · Electric double layer · Metal corrosion · Nernst equation · Sulphuration · Tin-free steel · Tinplate

Abbreviations

Al	Aluminium
AS	Anodic surface
CS	Cathodic surface
ECCS	Electrocoated chromium steel
i_{corr}	Corrosion current intensity
EMF	Electromotive force

© The Author(s) 2015
C. Barone et al., *Food Packaging Hygiene*, Chemistry of Foods,
DOI 10.1007/978-3-319-14827-4_6

E	Electrode potential
HCN	Hydrocyanic acid
H_2	Hydrogen
H_2S	Hydrogen sulphide
Fe	Iron
Mg	Magnesium
Mn	Manganese
η	Overvoltage
E°	Standard reduction potential
Sn	Tin
TFS	Tin-free steel
TP	Tinplate

6.1 Introduction

The term 'corrosion' is conventionally applied to the oxidation of a metal surface. The internal corrosion of food cans is characterised by the dissolution of the container metal (iron, tin, aluminium) in the packaged food. Foodstuffs react with the container, and the deterioration of metals occurs. The effect of this process is the declassification of food products to unmarketable articles because of organoleptic changes, vacuum loss, hydrogen swelling and metal concentration above the legal limit or perforation damages of the container.

The experience shows that the container 'lives' in perfect harmony with the content. The knowledge of involved mechanisms and the development of better materials and coatings have already reduced corrosion failures significantly in the last decades. In relation to the total production of food metal packaging, corrosion rates determine the diminution of the commercial shelf life in a very few cases. Nevertheless, the knowledge of the kinetic processes and correlated corrosion morphology is surely critical, including numerous and complex factors they are related to. The main aim is to manage failures (eradication or limitation) with adequate corrective and preventive actions. The control of metal corrosion in food packages is of great concern to packaging manufacturers, food processors and consumers.

Anyway, the final goal is to obtain and maintain the high qualitative level of canned foods by both sensorial and food safety viewpoints, reducing also nutritional variations if compared with the original product.

6.1.1 Basic Principles of Corrosion

In aqueous media, metal corrosion is an electrochemical process that involves the transfer of electrons. The electrolytic solution (food) is the medium transferring the electric current created by electron transfer. In this process, the metal surface acts as an electrode whose electron transfer equals the electronic transfer in the electrolytic medium.

In relation to electron exchanges, there is a reaction of metal oxidation by interaction with an environment that can be reduced. In this way, two reactions—6.1 and 6.2—occur simultaneously and complementary:

$$Me \rightarrow Me^{n+} + ne^- \tag{6.1}$$

$$R^{n+} + ne^- \rightarrow R \tag{6.2}$$

Reaction 6.1 corresponds to the anodic reaction or oxidation, while reaction 6.2 represents the cathodic reaction or reduction. These reactions are explained in detail in Sect. 6.1.2. Chemists define the corrosion reaction as 'oxidation–reduction'. Oxidation implies loss of electrons, whereas reduction means a pickup of electrons [1].

The metal that releases electrons leaves the crystalline structure and becomes a positive ion: it is called anode. The accumulation of electrons would establish a negative charge on the other ionised metal; its surface is protected from any dissolution and becomes the site where reduction reactions occur. This metal is called cathode. At any moment, cathodic current is equal to anodic current [2, 3].

The above-mentioned schematic model shows that a metal will only corrode in the presence of a cathode where ions can satisfy their tendency to absorb available electrons. Anyway, there is a physical or chemical heterogeneity as necessary condition of corrosion phenomena.

The general electrochemical scheme usually represents a corrosion process as follows:

(a) The 'shorted galvanic cell': anodic and cathodic areas—for example two different metals in contact—are macroscopically separated and
(b) The 'mixed' electrode: anodic and cathodic areas are not detectable.

From the physicist's viewpoint, there are not electrochemical processes that can produce absolutely homogeneous processed metals. In fact, microheterogeneity is permanent or temporarily present on the metal surface. For example, steel and aluminium—widely used as alloys—are always found to enclose microscopic dissymmetries: local anodes and cathodes [4].

When speaking of corrosion processes, two factors should be considered:

- The potential factor (thermodynamic aspects) and
- The facility factor (kinetic aspects).

6.1.2 Thermodynamic Condition of the Occurring of a Spontaneous Corrosion Process

The thermodynamic tendency of an electrode to oxidise or reduce may be expressed by means of the standard reduction potential (E°). This phenomenon of 'cationic transfer', which concerns a large number of metals with several exceptions, can be generally represented by means of reaction 6.1 (Sect. 6.1.1) where

- Me is the pure metal (anode).
- Me^{n+} is the positively charged metallic ion or cation, characterised by 'n' positive charges.
- ne^- represents 'n' electrons (negative charges).

When speaking of tin (Sn) and iron (Fe), the schematic reaction 6.1 can be substituted with reactions 6.3 and 6.4, respectively:

$$Sn \rightarrow Sn^{2+} + 2e^- \qquad (6.3)$$

$$Fe \rightarrow Fe^{2+} + 2e^- \qquad (6.4)$$

As a consequence of reactions 6.1, 6.3 and 6.4, metal cations migrate into the solution (electrolyte), while as many electrons remain on the metallic surface (electrode). The process continues until an equilibrium state between positive and negative charges, which appear in the right part of reactions 6.1, 6.3 and 6.4. In this way, an 'electric double layer'—approximately, an electrical capacitor with positive and negative charges—is obtained (Fig. 6.1).

The metal surface assumes a negative electric charge with respect to the electrolyte; it is characterised, i.e. by an electric potential of negative charge, called 'electrode potential' (E). However, it should be noted that not all metals send cations in an electrolyte. In fact, there is a number of metals—copper, mercury, silver,

Fig. 6.1 The electrode double layer. Tin cations migrate into the solution (electrolyte), while as many electrons remain on the metallic surface (electrode) and the process continues until an equilibrium state between positive and negative charges. An electric double layer is formed in this way

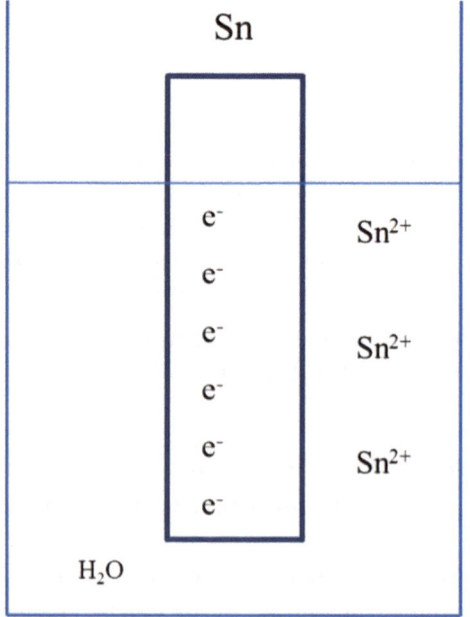

rhodium, palladium, iridium, platinum, gold—which are after the normal hydrogen in the series of the potential (this series is discussed later).

As a consequence, these metals have a lesser tendency to ionise; in fact, the metal surface assumes a positive electric charge with respect to the electrolyte. Anyway, the electrical double layer is formed even for above-mentioned metals. As an example, the order of magnitude of the thickness for an electric double layer is 10^{-7} cm [5].

Both cathodic and anodic reactions have their own reversible electrode potential in corrosion processes. The reversible potential of the anodic reaction (or oxidation) may be conventionally defined here E_{ox}, while the reversible potential of the cathodic reaction (or reduction) can be named E_{red}. After these premises, the theory predicts that the thermodynamic condition for a spontaneous galvanic process, such as a corrosion process, is expressed by means of Eq. 6.5:

$$E_{red} > E_{ox} \qquad (6.5)$$

The difference 'E_{red}—E_{ox}' is the electromotive force (EMF) of the corrosion process and defines the degree of tendency of a metal to release energy with a spontaneous corrosion. This process is spontaneous if EMF is >0, as shown in Eq. 6.6.

$$EMF = E_{red} - E_{ox} > 0 \qquad (6.6)$$

EMF depends on the relationship between E° and the standard Gibbs free energy, where the negative value represents the tendency of spontaneous reactivity under standard conditions ($\Delta G^\circ = -zFE^\circ$).

Corrosion phenomena of metals can be well studied on condition that the so-called series of standard potential is known. This series, also named 'series of standard potential' or 'electrochemical series' (Table 6.1), indicates the standard potential of a fairly large number of metals according to increasing values in volts. Electrode potentials are defined in connection with the potential of a reference electrode: the hydrogen electrode, made up of a platinum wire in an acid solution at unitary concentration on which hydrogen is bubbled at a pressure of 1 atm.

According to Nernst, the reversible or equilibrium potential of a metal electrode dipped in a solution of its salt at a concentration different from the unitary (1 M solution) can be calculated again with respect to the normal hydrogen electrode as shown by Eq. 6.7:

$$E_{rev} = E^\circ + \frac{RT}{nF} \text{Ln} \left(Me^{n+} \right) \qquad (6.7)$$

where

(Me^{n+})	ionic concentration of the solution of a salt of the metal Me dipped in the same solution
E^0	normal potential of Me
R	universal gas constant = 8.315 J/(K × mol) = 8.315 V C/(K × mol)
T	absolute temperature, K
F	Faraday's constant: 96,500 C/equivalent
n	oxidation number (valence) of Me

Table 6.1 Standard EMF series of metals [5]

Metal–metal ions equilibrium (unit activity)	Electrode potential versus normal hydrogen electrode at 25 °C (V)	Metal–metal ions equilibrium (unit activity)	Electrode potential versus normal hydrogen electrode at 25 °C (V)
Li/Li$^+$	−3.045	V/V^{+++}	−0.876
Rb/Rb$^+$	−2.925	Zn/Zn^{++}	−0.762
K/K$^+$	−2.925	Cr/Cr^{+++}	−0.740
Cs/Cs$^+$	−2.923	Ga/Ga^{++}	−0.530
Ra/Ra$^+$	−2.920	Fe/Fe^{++}	−0.440
Ba/Ba^{++}	−2.900	Cd/Cd^{++}	−0.402
Sr/Sr^{++}	−2.890	In/In^{++}	−0.342
Ca/Ca^{++}	−2.870	Tl/Tl$^+$	−0.336
Na/Na$^+$	−2.714	Mn/Mn^{+++}	−0.283
La/La^{+++}	−2.520	Co/Co^{++}	−0.277
Mg/Mg^{++}	−2.370	Ni/Ni^{++}	−0.250
Am/Am^{+++}	−2.320	Mo/Mo^{+++}	−0.200
Pu/Pu^{+++}	−2.070	Ge/Ge^{++++}	−0.150
Th/Th^{++++}	−1.900	Sn/Sn^{++}	−0.136
Np/Np^{+++}	−1.860	Pb/Pb^{++}	−0.126
Be/Be^{++}	−1.850	Fe/Fe^{+++}	−0.036
U/U^{+++}	−1.800	H$_2$/H$^+$	0.000
Hf/Hf^{++++}	−1.700	Cu/Cu^{++}	+ 0.337
Al/Al^{+++}	−1.660	Cu/Cu$^+$	+ 0.521
Ti/Ti^{++}	−1.630	Hg/Hg^{++}	+ 0.789
Zr/Zr^{++++}	−1.530	Ag/Ag$^+$	+ 0.799
U/U^{++++}	−1.50	Rh/Rh^{+++}	+ 0.800
Np/Np^{++++}	−1.354	Hg/Hg^{++}	+ 0.857
Pu/Pu^{++++}	−1.280	Pd/Pf^{++}	+ 0.987
Ti/Ti^{+++}	−1.210	Ir/Ir^{+++}	+ 1.000
V/V^{++}	−1.180	Pt/Pt^{++}	+ 1.190
Mn/Mn^{++}	−1.180	Au/Au^{+++}	+ 1.500
Nb/Nb^{+++}	−1.100	Au/Au$^+$	+ 1.680
Cr/Cr^{++}	−0.913		

The EMF series can be seen as a sort of list of different metals on the basis of the standard oxidation–reduction potentials. Basically, most electrochemically active metals are remarkable negative standard potentials. On the other hand, electrochemically 'inert' metals tend to be reduced negative standard potentials. Practically, a couple of two metals can be seen with an anode (the most active metal) and a cathode (the 'inert' metal). The first metal (anode) is able to corrode

The Nernst equation can be also expressed by Eqs. 6.8 and 6.9:

$$E_{rev} = E° + \frac{0.0596}{n} Log(Me^{n+}) \tag{6.8}$$

$$E_{rev} = E° + \frac{RT}{nF} Ln\left(\frac{Ox}{Red}\right) \tag{6.9}$$

6.1.3 Kinetic Aspects of the Corrosion Processes: Polarisation Phenomena

In a corrosion process of a metal, as has been previously explained, Eqs. 6.5 and 6.6 do not give information on the 'rate' of the corrosion process and the correlated evolution over time (in kinetic terms). Electrode potentials are equilibrium potentials. In relation to the evolution of corrosion processes, it should be recalled that polarisation phenomena may be defined such as the moving of the reversible (equilibrium) potential of the anodic reaction in the positive direction (ηa) and of the cathodic reaction in the negative direction (ηc). A part of the available driving force is dissipated as polarisation or overvoltage, η. Equations 6.10 and 6.11 show the relation between η, E and the corrosion potential (E_{corr}).

$$\eta a = E - E_{corr} \tag{6.10}$$

$$\eta c = E_{corr} - E \tag{6.11}$$

Consequently, the type and the intensity of polarisation phenomena determine the rate of possible corrosive phenomena.

The flow of electrons from the anode to the cathode area makes the cathode increasingly less positive. In other words, the cathode undergoes cathodic polarisation. At the same time, the removal of electrons from the anodic area makes the anode increasingly less negative. In other terms, the anode undergoes anodic polarisation.

As a result, two polarisation curves are obtained. One of these curves is defined anodic polarisation curve, while the second of these functions is named other cathodic polarisation curve. These mathematical functions represent, therefore, the kinetic aspect of a corrosion process [6]. The value of the current intensity corresponding to the intersection of the two polarisation curves [6] is defined 'corrosion current intensity' (i_{corr}). Additionally, the potential corresponding to the same intersection of the two curves is named 'corrosion potential'.

The corrosion potential is an equilibrium potential between E_{ox} and E_{red}, while i_{corr} is directly correlated with the 'rate of the corrosion process'.

The overtension shows the difficulty of electron transfer under a given corrosion current intensity. As a result, thermodynamic and kinetic aspects have to be taken into account when evaluating a corrosive phenomenon: a possible corrosion process may have no practical consequences because its rate is close to zero. In relation to metal containers, a corrosion rate is acceptable even if different from zero, but the preservation of packaged products throughout its shelf life has to be guaranteed.

6.2 The Metal Packaging

Metal containers are widespread in the food industry due to their unique characteristics of mechanical strength and impermeability to gases and light, allowing long commercial shelf-life values.

Used materials in the manufacture of metal packaging are essentially three types: tinplate, tin-free steel (TFS) and aluminium alloys. Cans and ends are manufactured starting from very thin sheets (0.09–0.25 mm). Cans are produced as two-piece—or three-piece—structures, while ends can be defined as 'open top', 'easy open' and 'easy peel'.

Metallic materials may be protected with an organic coating of different nature. TFS and aluminium are used always lacquered, while can bodies in tinplate can also be used without coating.

Tinplate [7] is a heterogeneous material, defined in the Euronorm 10202:2004 [8] as 'sheet or roll of steel with a low carbon coated on both sides of the tin coating applied by continuous electrolytic deposition'. A section of its complex structure is represented in Fig. 6.2. According to the Euronorm 10202:2004, the tin coating weight may range from 2.0 to 11.2 g/m^2 per side (Table 6.2).

The same Euronorm defines TFS [9] as 'sheet or roll of steel with a low carbon coated on both sides by means of continuous electrolytic deposition of a coating composed of metallic chromium covered by an upper layer of chromium oxide' (Table 6.3).

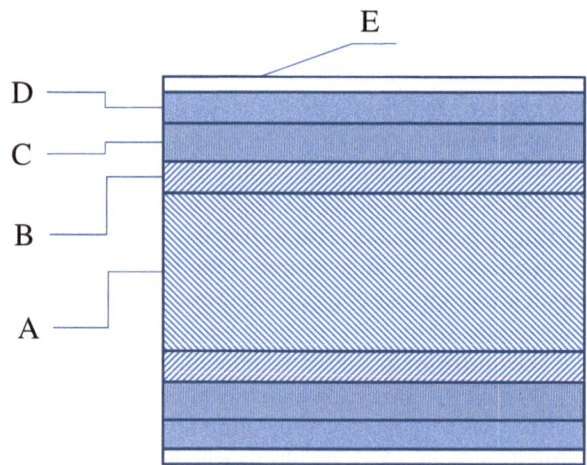

Fig. 6.2 Schematic structure of tinplate materials. The *A* layer is essentially steel base (0.15–0.49 mm). The *B* layer represents iron–tin alloys (0.1 μm) on both sides, while *C* and *D* layers are for free tin (0.25–1 μm) and passivation film (0.02 μm), respectively. Finally, a protective *E* layer of food-grade oil (0.0005 μm) is placed

Table 6.2 Nominal tin coating weight on tinplate according to UNI EN 10202:2004 [8]

Tinplate code	Nominal tin coating weight (g/m^2)	
	Side I	Side II
E 2.8/2.8	2.8	2.8
D 5.6/2.8	5.6	2.8
D 8.4/2.8	8.4	2.8
D 8.4/5.6	8.4	5.6
D 11.2/2.8	11.2	2.8
D 11.2/5.6	11.2	5.6

Table 6.3 Nominal chromium coating weight according to UNI EN 10202:2004 [8, 9]

Nominal chromium coating weight for each side	Minimum value (mg/m^2)	Maximum value (mg/m^2)
Total chromium	50	140
Chromium oxide	7	35

The total chromium is the sum of the content of metallic chromium, chromium oxide and hydroxide

Table 6.4 Commercially available types of aluminium alloys

Alloy	Composition	Food packaging applications
1070	Aluminium 99.7 %	Semi-rigid containers, easy-peel ends
3105	Al + Mn 0.5 % + Mg 0.5 %	Food and beverage cans
3004	Al + Mn 1.2 % + Mg 1.0 %	Food and beverage cans
5182	Al + Mn 4.5 %	Rings for easy-open ends

Aluminium used for the production of can bodies and ends is always made [10] of a three-component alloy—aluminium (Al), manganese (Mn) and magnesium (Mg)—in different ratios (Table 6.4), in order to improve mechanical features.

6.2.1 Internal Corrosion of Metal Packages

Foodstuffs packed in metallic cans are complex systems with different pH, buffer power and chemical compositions. These factors can either accelerate or inhibit the corrosion and influence correlated mechanisms. In fact, the corrosion of metal packs can show various morphologies and follow several mechanisms. First of all, corrosion phenomena depend on the type of metal material: tinplate, TFS or aluminium. Secondly, corrosion is affected by the presence of an organic coating. The corrosion resistance is also influenced [11] by several factors related to:

- Packaging features (monometallic or bimetallic, plain or lacquered material)
- Food composition
- Filling process
- Storage conditions.

The metallic packaging is a closed system, without any exchange with the external environment: this fundamental factor has to be considered when examining different corrosion mechanisms.

6.3 Tinplate

The corrosion of tinplate is a more complex phenomenon if compared to the previously described process. In detail, the corrosive process can be considered as the result of the concurrence of several cathodic and anodic subprocesses that elapse at the interphase of a polyelectrode at a common potential, the mixed potential of corrosion.

Actually, the surface of a tinplate can is a very heterogeneous electrode whose different layers, due to raw materials, are shown with superficial ratios, which depend on productive conditions.

As regards the field of the fundamental reactions of corrosion of the tinplate, two schemes can be usefully considered. The first of these schemes refers to the electrochemical coupling of Sn and Fe: a galvanic cell (Fig. 6.3). Exchanged currents among different anodic and cathodic areas can affect a vast surface of the electrolytic conductor; consequently, there is a highly marked influence of the conductivity of the environment and a predominant importance of superficial geometrical factors.

The second scheme concerns a real coupling of two metals (Sn and Fe are in contact); therefore, the anodic reaction on Sn and the cathodic reaction on the microzones of the uncovered steel are expected. As a result, different corrosion mechanisms can be developed; they are summarised in Fig. 6.4.

It has to be noted that the inner side of cans is lacquered [12] when the general purpose is to limit phenomena of interaction between tinplate and canned foods. These foods have medium or high acidity or contain sulphur compounds. Lacquered cans may be also preferred for aesthetic reasons.

In general, two typical cases can be distinguished with concern to the preservation (integrity) of lacquer films in every step of can manufacturing:

- Shallow discontinuities (holes, small scratches, abrasions) which only affect the paint film and the tin coating
- Deep discontinuities (scratches, cuts) which affect also steel.

Fig. 6.3 Electrochemical coupling of tin and iron. Schematic structure of a bimetallic (Sn/Fe) galvanic macroelement

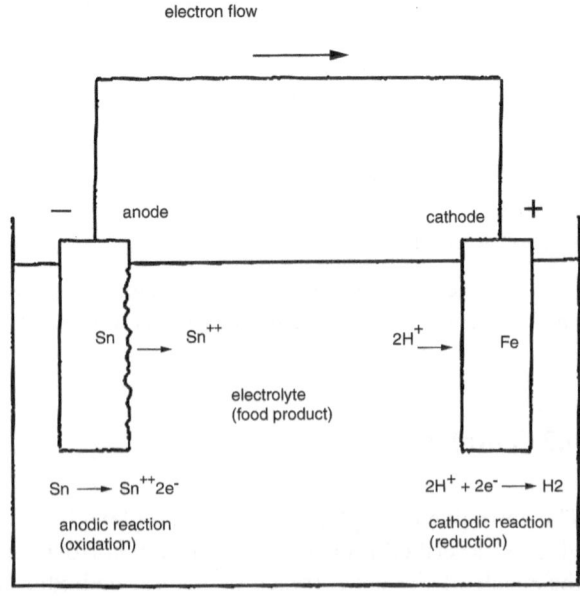

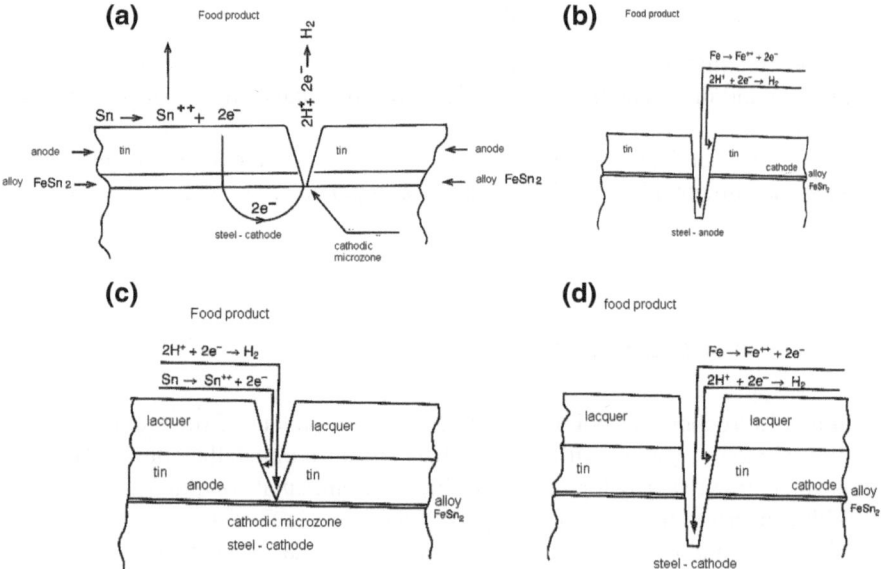

Fig. 6.4 Scheme of different types of corrosion mechanisms on tinplate materials. **a** Tin as anode; **b** steel as anode; **c** lacquered tinplate, tin as anode; and **d** lacquered tinplate, steel as anode

This discussion is particularly interesting because corrosive phenomena usually take place in correspondence of coating discontinuities of the film itself (holes, abrasions, fractures). The type of discontinuity has a peculiar importance with concern to the kinetics of corrosion processes, because of the modified relationships between anodic and cathodic areas.

The presence of lacquer [13] has the main effect of changing these superficial ratios. In fact, all metallic materials of the can are electronically in contact with each other and with the foodstuff, even with a different relative surface. Consequently, a polyelectrode is formed at the liquid–solid interphase because of the interaction of several galvanic couples.

The nature of single interphases and, hence, the relative electrochemical behaviour of different metal components depend on the physical and chemical characteristics of the material and the electrolyte.

Canned foods are generally distinguished, according to their level of aggressiveness, in:

- Non-aggressive products; absence of aqueous phase (e.g. dried fruit, pasta, powdered products)
- Medium aggressive foods; medium or acid pH due to the presence of organic acids such as citric acid (e.g. derivatives of tomato, fruit in syrup)
- Highly aggressive products; acid pH due to the presence of organic acids such as acetic acid (e.g. pickled pearl onions, *sauerkraut*)

- Sulphurs; products containing sulphur proteins (e.g. tuna, meat, *pâté*).

This macrodistinction can be modified in turn by different factors depending on the product and/or its conditions of preparation (residuals antiparasites, cold packaging) [14].

The electrochemical behaviour of tinplate depends on the aggressiveness of the product and on superficial ratios of metal components, as described in the next sections.

6.3.1 Tin as the Anode in the Tin–Iron Couple

From an electrochemical point of view, the Sn component of tinplate materials is an electrode that, together with the steel base, forms a bimetallic couple. This couple is vulnerable to corrosion under specific environmental conditions.

With concern to the accurate analysis of the behaviour of the Sn/Fe couple, the position of tin and iron in the electrochemical series has to be necessarily considered. $E°$ values are -0.136 and -0.440 V with reference to Sn/Sn^{2+} and Fe/Fe^{2+} reactions, respectively.

The standard potential of Sn is less negative if compared with $E°$ for Fe; consequently, tin must assume the cathodic role in the Sn/Fe couple. Actually, Sn is also reported to assume a negative potential compared to iron (anode) when speaking of medium/acid canned foods containing citric acid (out of air contact). Main causes are thermodynamic factors (formation of stable complexes with some organic acids) and kinetic reasons (high hydrogen overvoltage of tin). Consequently, Sn corrodes preferentially (detinning phenomena) with an effect of protection from the corrosion of steel base.

A very big anodic surface (AS) and a remarkably small cathodic surface (CS) are needed, so that Sn can protect Fe from corrosion in the above-mentioned conditions. Therefore, the anode is represented by Sn (metallic coating) in the tinplate and must be uniform and covered as much as possible. On the other hand, the cathode is formed by small discontinuities (holes, abrasions) with uncovered steel: the global extension of discontinuities means a very small CS. In this way, the cathodic protection of steel can be observed (Fig. 6.4). Mathematically, a new conceptual idea—the coexistence of a big anode and a small cathode—can be expressed with the big ratio S_A/S_C between anodic (S_A) and cathodic (S_C) areas, and the consequent low corrosion rate.

6.3.1.1 Fundamental Reactions

With concern to products that are predominantly detinning (fruits and vegetables with medium and low acidity) in anaerobic conditions and packaged into tinplate cans, the fundamental reactions of corrosion are as follows:

- Attack and solubilisation of tin (anode) to give Sn^{2+} ions which, migrating into the canned food, are 'complexed' by several substances (organic compounds such as citric, malic and tartaric acids). As a result, the concentration of tin ions in the solution remains sufficiently low and the reversible potential is practically constant. Reaction 6.3 shows the anodic reaction or oxidation.
- Discharge of H^+ ions coming from the acidic substances in canned foods on the cathodic zones, with the consequent formation of gaseous hydrogen. The cathodic reaction or reduction is displayed by reaction 6.12.

$$2H^+ + 2e^- \rightarrow 2H \rightarrow H_2 \qquad (6.12)$$

The current generated by the reduction of residual oxygen is added to the cathodic current in the first hours after the packaging (reaction 6.13).

$$O_2 + 4H^+ + 4e^- \rightarrow 2H_2O \qquad (6.13)$$

The anodic role of Sn is also due to its high overvoltage of hydrogen compared to overvoltage values of iron and the $FeSn_2$ alloy (placed between tin and steel). As a result, the discharge of H^+ ions takes place on steel instead of Sn.

The reaction of the corrosion of tin, also named 'detinning' reaction, can be expressed by reaction 6.14 from the combination of relations 6.3 and 6.12:

$$Sn + 2H^+ \rightarrow Sn^{2+} + H_2 \qquad (6.14)$$

In real packages, the electrochemical behaviour can be more complex as the corrosive process and the corrosion rate depend on the superficial ratio of different metals. Moreover, several anodic and cathodic reactions can occur simultaneously to the process and elapse at a common potential (the mixed corrosion potential) to which the sum of cathodic currents equals the sum of anodic currents.

Based on relation 6.14, the edible content that is created within cans with plain body is a reducing environment due to the presence of hydrogen. This environment is very important for some canned products (colour maintenance for the fruit with white flesh).

As regards the prevailing cathodic reaction 6.12, namely the formation of atomic and finally molecular hydrogen (H_2), it must be considered that H_2, although being in quantities stoichiometrically proportional to tin migrated in the canned food, partly spreads from the inner walls of the can outwards through the network of the ferrite (α Fe), which is the component of the ferrous matrix of the steel in the tinplate.

The kinetics of corrosion of aggressive canned foods in plain containers can be represented as shown in Table 6.5 with a peculiar sequence. In relation to lacquered cans, the corrosion starts in correspondence of a hole or other type of lacquer discontinuity. Subsequently, corrosion goes on 'under skin' to the coating–tin interface, with a possible lifting of the film (also for very limited surfaces) and the darkening of uncovered zones (Fig. 6.4).

Table 6.5 Kinetics of tinplate (TP) corrosion on different materials

Corrosion steps	Aggressive/corrosive food product for tin		Aggressive food product for tin	
	Plain TP	Lacquered TP	Plain TP	Lacquered TP
First period (few days)	Fast corrosion on tin and steel	Slow corrosion on tin and steel	Fast corrosion on steel	Fast corrosion on steel
Second period (some months or year)	Slow corrosion on tin	Undermining corrosion on tin	Steady corrosion on steel	Steady corrosion on steel
Third (short period)	Corrosion on tin and steel	Corrosion on steel		
Failures	Hydrogen swelling	Hydrogen swelling	Hydrogen swelling	Perforation
	Tin over the legal limit in the product	Blistering	Perforation	Hydrogen swelling
		Lacquer detachment	Steel over the legal limit in the product	Steel over the legal limit in the product

The description of corrosion steps can vary depending on the nature of canned foods (aggressive and/or aggressive and corrosive product) and the presence of coating layers on tinplate

The iron of the steel base remains protected cathodically by Sn. The second period of corrosion can be shorter if compared to a plain can, especially when the presence of deep discontinuities up to the steel base can be observed. In fact, the critical surface relationship is reached more rapidly; therefore, steel is no longer protected by tin (Table 6.5). This phenomenon, also known as 'undermining corrosion', can have several origins:

- Non-uniform adherence of the lacquer to the tinplate, that might be due to non-uniform oil or passivation film on the tinplate, or
- Fragile lacquer films with possible fractures because of subsequent mechanic working procedures on tinplate surfaces.

Moreover, oxygen (dissolved in canned food) can contribute to the weakening and to the detachment of the film (presence of hydroxide anions) because it is reduced to the lacquer–tinplate interface.

In summary, as regards the undermining corrosion, the reactions 6.3 and 6.4 (anodic reactions) can be considered with reaction 6.15 (cathodic reaction):

$$O_2 + 2H_2O \rightarrow 4\,OH^- + 4e^- \tag{6.15}$$

The kinetics of corrosion of coated cans is shown in Table 6.5. The second period of corrosion can be shorter if compared to a plain tinplate can, especially if there are deep scratches. Should this be the situation, the package would be subject to fail due to swelling defects, while the exceeding tin in solution does not appear to be the main cause.

6.3.2 Iron as Anode in the Tin–Iron Couple

As shown in Sect. 6.3.1, the normal potential of Sn is less negative than that of Fe; if there are no reactions lowering the potential of tin, or if there are reactions modifying the potential of iron, Fe takes the role of anode in the couple Sn/Fe.

This situation happens when speaking of plain tinplate (out of the contact with air), and there are substances that activate specifically the corrosion of steel or inhibit corrosive processes on Sn. Consequently, the corrosion develops at the depth of small uncovered areas of iron. Tin does not corrode, and it acts as cathode (Fig. 6.4).

6.3.2.1 Fundamental Reactions

The fundamental reactions of corrosion in tinplate cans containing canned predominantly aggressive foods for steel in anaerobic conditions are the following ones:

- Attack and solubilisation of Fe (anode) to give Fe^{2+} ions
- Discharge of H^+ ions on cathodic zones coming from acid substances present in the canned food, with development of gaseous hydrogen.

Anodic areas (steel) are small, and surrounding cathodic areas (tin) are vast. The ratio of S_C to S_A is very big, differently from the analogue ratio of Sect. 6.3.1. This condition favours the intensity of very high-localised i_{corr} current and lays the premise for can failures, due to perforations of the can itself. This phenomenon represents the last step of the particular process of corrosion, also known as 'pitting corrosion'.

As regards the cathodic reaction, it has to be necessarily noted that formed H_2 cannot diffuse outside of the container due to the crystalline structure (compact tetragonal) of Sn. Consequently, hydrogen gathers within the container causing swelling. From a practical viewpoint, this situation is only theoretical and can be verified only by means of anomalies in the composition of the food product such as pears in syrup [15]. All packages of products that are aggressive for steel are internally lacquered.

As regards lacquered cans, the further localisation of the corrosive attack near some painting pores increases the risk of premature failure of the packages due to swelling caused by hydrogen or perforation (Fig. 6.4). Basically, the layer of tin is almost entirely protected, and the attack of steel proceeds deeply; the superficial relationship between steel and Sn does not vary considerably over time, and the corrosion rate is practically constant.

Finally, a general observation must be made: above-described corrosive phenomena can also cause alteration in the taste and colour of the canned food. The kinetics of corrosion is described in Table 6.5.

6.3.3 Morphological Aspects of the Internal Corrosion of Plain Tinplate Cans

Main morphologies of corrosion can have the following features, depending on canned products, the elaboration of products and storage periods:

- Slight to highly intense detinning, which can only affect tin or the iron–steel alloy until basic steel (Fig. 6.5)
- Pitting and deep craters of small dimensions, which can affect all the thickness of the material (Fig. 6.6).

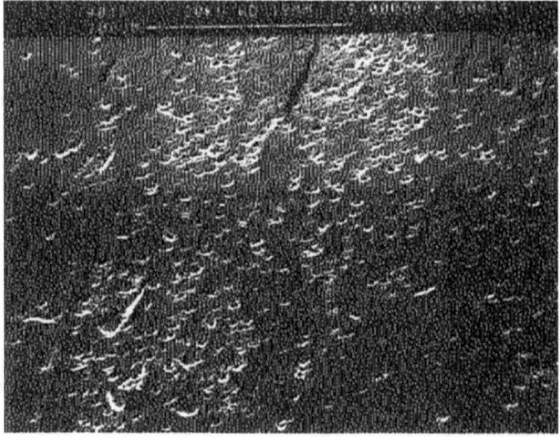

Fig. 6.5 Morphological aspects of corrosion on tinplate. Light detinning effects that can only affect tin or the iron–steel alloy until basic steel [23]

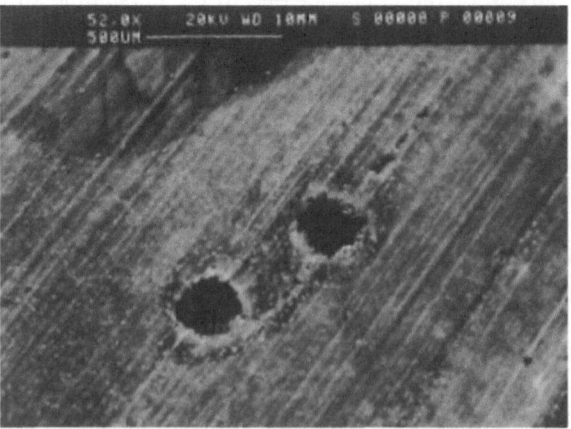

Fig. 6.6 Morphological aspects of corrosion on plain tinplate. Pitting and craters of small dimensions developed in depth, which can completely affect the thickness of the material [23]

6.3.4 Morphological Aspects of the Corrosion of Cans with Lacquered Body and Can Ends

In relation to cans with lacquered body, main morphologies of corrosion can be (depending on the type of canned products, the food elaboration and storage periods) as follows:

- Points of corrosion
- Perforations
- Sulphurations on holes, scratches and abrasions (this defect occurs only with tinplate materials when speaking of bottoms and ends)
- Total or partial lack of adherence
- Undermining corrosions or black spots, detachment or blistering of the lacquer (Fig. 6.7).

6.3.5 Variables Influencing Tinplate Corrosion

6.3.5.1 Influence of the Product

The aggressiveness of canned food towards tinplate surfaces depends on the nature of main components (including possible residuals of antiparasitic treatments) [16], on the use of several ingredients in food preparation and on the packaging technology. Among edible components of canned foods, some substances act as corrosion accelerators carrying out a prevalent action of anodic or cathodic depolarisation.

From a general viewpoint, food products with pH > 5.0–5.5 are not cause of corrosion of the unlacquered tinplate.

As regards pH values lower than 5.0 and particularly for foods with pH between 3.0 and 4.5, corrosion rates become more rapid with pH decrease.

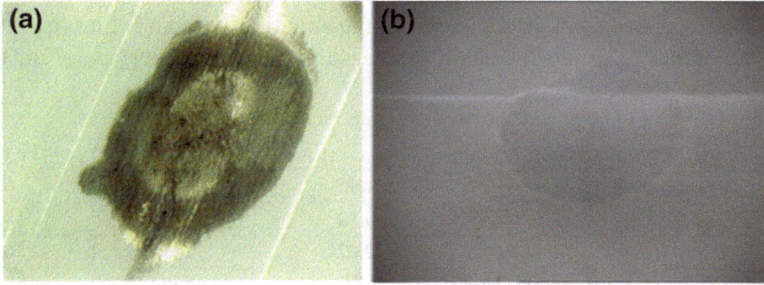

Fig. 6.7 Main morphologies of undermining corrosion of cans with lacquered body. The corrosion starts in correspondence with a real (**a**) or potential (**b**) hole of the lacquer and proceeds under the organic coating, with detachment

Organic acids are 'complexing' natural substances contained in several products, particularly fruit and vegetables; a typical example is the 'white' fruit. Organic acids can produce complexes with Sn^{2+} cations that pass in solution, following the attack of tinplate supports. Organic tin compounds are obtained in this way, where Sn is part of the molecule with the complexing substance. Among the most important substances forming complexes with tin, the following compounds can be mentioned: tartaric acid, malic acid, citric acid and oxalic acid, in increasing order of complexing power. The influence of the complexing power of several organic acids with Sn can be efficacy demonstrated by means of data shown in Table 6.6. Other components of foods such as flavonoids and anthocyanins have complexing power towards Sn.

The complexing agents of iron are natural substances that form complexes with Fe^{2+} ions: these cations flow into the solution following a corrosive preferential attack on the steel base. Therefore, they act as cathodic depolarising agents. The most well-known Fe-complexing agents are as follows:

- Rutin. It is already mentioned as anodic activator, and it can form black-coloured complexes with Fe^{2+}.
- Amygdalin. This substance is a glycoside that can be found in bitter almonds and in the pit of fruits. It is responsible for swelling caused by hydrogen in cans containing non-pitted fruit and attributed to hydrocyanic acid (HCN) that is enzymatically formed by means of the hydrolysis of amygdalin (with β-glucosidase). HCN is a strong Fe^{2+}-complexing agent; this attitude also explains the attack of steel base and the rapid swelling by hydrogen.
- Tannins. Mushrooms, artichokes, asparagus and chestnut sauce contain tannins, which form complexes such as ferric tannates with a bluish-black colour.

As regards nitrates and oxygen, great accelerators of corrosion, these molecules are responsible for a concurrent cathodic process with respect to metal oxidation.

Oxygen is an activator of corrosion towards both tin and steel. Steel can be scarcely protected by tin in aerated conditions or even become the preferential anode of the couple in the presence of Sn-complexing agents.

Moreover, the harmfulness of the corrosive action of oxygen is exacerbated by durable effects over time (when oxygen has been eliminated). The corrosion proceeds at higher speeds if compared to the situation without initial oxygen: this phenomenon is due to the substantial increase in the relative surface of uncovered

Table 6.6 Influence of complexing power of organic acids with tin

Electrolyte with Sn-complexing attitude	I_{corr} ($\mu A/cm^2$)	Sn (mg/kg)	Fe (mg/kg)
Lactic acid (2 g/l)	1.45	58.9	6.7
Acetic acid (0.5 g/l)	0.12	0.0	14.4
Salt + citric acid (2 g/l)	0.024	141.0	0.7
Salt + citric acid (1.3 g/l) + lactic acid (2 g/l) + acetic acid (0.7 g/l)	7.00	535.0	155.0

Organic tin compounds are formed in this way, where Sn is part of the molecule of the complexing substance steel during the corrosion in the presence of oxygen (reaction 6.13). For these reasons, the reduction of air levels and oxygen in a tinplate can must be carried out through adequate precautions when filling and closing packages.

Anyway, traces of oxygen can be mostly localised in the headspace even after well-managed canning and closure processes. The residual oxygen causes the typical phenomenon of corrosion known as 'water line' attack. These words mean a ring of corrosion, sometimes accentuated, on the walls of can bodies near headspace boundaries. This phenomenon is known as corrosion by differential aeration.

Nitrates are found in plants, water supplies and heavily fertilised soils [14]. Further studies have demonstrated that green beans, spinach, lettuce and basil contain several thousands mg/kg of nitrates. During storage, nitrates are reduced through a number of intermediates (nitrite ions) to ammonia and act as corrosion accelerators [17], giving rise to a cathodic process completing the concomitant hydrogen reduction.

Table 6.7 shows and compares the influence of the residual oxygen (obtained through the variation of net weight), nitrates and pH in canned tomato paste, packed in plain cans after 36 months of storage at room temperature. The product has been filled at a temperature of 50 °C with a sugar content of 7 °Bx and the initial content of nitrates of 18 mg/kg [18].

This research has demonstrated that pH has a negligible influence on the detinning process, while the influence of nitrates is decisively important: 10 mg/kg of nitrates causes the corrosion of 50 mg/kg of Sn, confirming the role of corrosion activators.

The same trial carried out with lacquered cans has confirmed the role of nitrates. These anions accelerate the corrosion rate because of their detinning

Table 6.7 Effect of nitrate concentration and other parameters on detinning corrosion

Constant parameters	Variables	Amount level	Average Sn concentration (mg/kg)	Detinning effect as difference (Δ) between initial and final amounts of Sn (mg/kg)
Tomato paste (natural nitrates: 18 mg/kg)	Net weight	400 g	222	
		410 g	197	$\Delta = 25$
		420 g	179	$\Delta = 18$
Filling temperature: 50 °C	Initial nitrate amount	Initial value + 20 mg/kg	253	$\Delta = 58$
		Initial value + 10 mg/kg	195	$\Delta = 45$
		Initial value	150	
Sugar content: 7 °Bx	pH value	4.15	204	$\Delta = 9$
		4.40	195	$\Delta = 9$

Canned food Tomato paste

Table 6.8 Effect of nitrates and other elements on corrosion

Presence of corrosive conditions	20 °C after 36 months	37 °C after 36 months	50 °C after 18 months
Nitrates	No acceleration	Acceleration	Acceleration
Cuprum	No acceleration	No acceleration	No acceleration
Nitrates + cuprum	No acceleration	No acceleration	No acceleration
Nitrates + scratch (the worst condition)	Acceleration	Acceleration	Acceleration
Cuprum + scratch	No acceleration	Acceleration	No acceleration

Acceleration phenomena in canned tomato paste (Table 6.7) with lacquered cans

power. In fact, the worst condition has been observed when an additional amount of nitrates has been recorded (Table 6.8). At low temperature, the presence of a scratch is also required to make possible the action of nitrates; undermining corrosion takes place at higher temperatures.

6.3.5.2 Influence of the Metallic Material

The chemical composition of steels intended for the production of tinplate influences the corrosion resistance [19]. In particular, the following parameters play an important role:

- Structure of steels, depending on the composition and both on cold and hot metallurgical processes; the presence of non-metallic inclusions (Al_2O_3) influences perforating corrosion
- Conditions of the surface
- Presence of metalloids and metals in the composition of steel.

The optimisation of the first two parametric groups (structure and surface) allows obtaining tinplate with good coating uniformity and superficial quality. These features contribute considerably to the achievement of good characteristics of corrosion resistance. Some metalloids and metals have, instead, the role of cathodic depolarisers, such as:

- Sulphur. Sulphurs contained in the steel base of tinplate have an unfavourable influence on the corrosion resistance in detinning environments because they act as cathodic depolarisers.
- Phosphor. It acts as a cathodic depolariser, but only in detinning conditions.
- Copper. It influences considerably the corrosion as cathodic depolariser when the related amount is higher than 0.06 %.

The influence of steel composition on the corrosion rate is clearly demonstrated on the basis of the comparison of data reported in Table 6.9, where D8.4 tinplate produced from L-type steel (cleaner) is compared to D8.4 tinplate produced from MR-type steel, with a higher content of phosphorus.

Table 6.9 Influence of the composition of L or MR steel base (D8.4 tinplate) on corrosion rates

Sample	Net weight (g)	Vacuum degree (mmHg)	Iron (mg/kg)	Tin (mg/kg)
Time = 28 months at 20 °C				
E3MR	408.0	298.5	9.6	84.8
E4L	409.8	362.0	4.8	62.2
E3MR	414.3	165.1	7.9	90.0
E4L	409.4	260.8	10.3	45.9

The concentration of iron and tin is higher for foods packed in MR-type tinplate. The difference of behaviour between two types of steel is evident even in the presence of corrosion accelerators such as nitrates.

6.3.5.3 Influence of Filling Conditions

As above explained, the reduction of air level and oxygen in tinplate cans is a crucial factor for shelf life; adequate interventions during the filling and closure of the package are needed. The main technologies of packaging for food metal containers are shown in Table 6.10.

Several devices for the reduction of residual air are available at present: the 'hot filling', the 'steam jet', the prefilling systems and the use of modified atmosphere. The type of measures depends on the type of the product and on processing conditions.

6.3.5.4 Influence of the Conditions of Storage and Transport

After the thermal treatment of stabilisation, the metal packaging is subject to the rather complex logistical activity of transport, storage and distribution up to the final consumer. During this period, some external events can influence corrosive processes and, therefore, the commercial life of the product. Among these events, mechanical (hits and vibrations) and climatic (temperature and humidity) factors have to be taken into account.

Table 6.10 Packaging technology of canned foods

Packaging technology of canned food in rigid containers	
Type technology	Examples
Hot packaging in one-stage single phase; T > 85 °C	Tomato paste
Packaging in two phases: solid + liquid	Diced vegetables and fruits
Vacuum-sealed packaging (5–6 cmHg)	Legumes: chickpeas and peas
Modified atmosphere packaging	Carbon dioxide and nitrogen (coffee)
Aseptic packaging	Glass and metal containers

Mechanical damage and transport vibrations under load can accelerate corrosive phenomena both modifying the superficial relationships and creating the conditions for the activation of a localised corrosion process such as stress corrosion cracking [20]. Environmental humidity is responsible for corrosive phenomena of the external surface of containers. Finally, the temperature is the main factor to be considered because it can act both directly and indirectly favouring the formation of condensation on external surfaces.

In accordance with Arrhenius law, the speed of a chemical reaction increases exponentially as temperature rises, taking into account the 'activation energy'. This peculiar parameter enables to determine the factor of acceleration of rate or temperature coefficient, Q10. Electrochemical reactions of corrosion are also highly influenced by temperature [2]. For these reasons, the control of temperature is fundamental during the commercial shelf life, in particular:

- Cooling temperature
- Storage temperature.

A Q10 factor of about two can be assigned to the detinning phenomenon when the can is internally plain. On the other hand, Q10 can also reach values of 3–4, for example in tomato products canned in varnished cans.

Tables 6.11 and 6.12 concern the influence of storage and cooling temperatures, respectively, on the corrosion of plain cans, in terms of dissolved tin. In addition, Fig. 6.8 is correlated with the influence of storage temperatures on iron corrosion. In summary, corrosion intensity appears to increase when storage temperatures increase and cooling temperatures slow down.

The development of corrosion in a lacquered can is shown by the concentration of iron; the influence of storage temperature on packaging corrosion (diced tomato) is clearly visible in Fig. 6.8.

Table 6.11 The influence of storage on the corrosion of plain cans, in terms of dissolved tin

Temperature (°C)	Plain can
	Tin concentration (mg/kg) after 3 months storage
20	40
37	70
55	98

Table 6.12 The influence of cooling temperatures on the corrosion of plain cans, in terms of dissolved tin

Cooling rapidity	Dissolved Tin (mg/kg)
Fast	40
Slow	80

Fig. 6.8 Influence of storage temperatures on iron corrosion (canned diced tomato in lacquered metal cans)

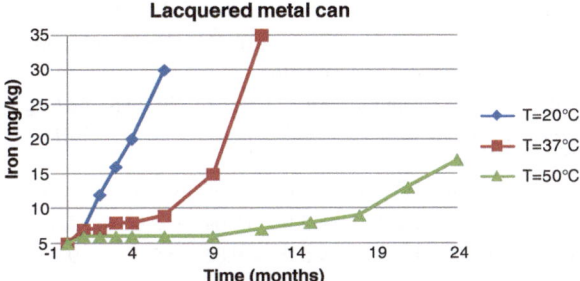

6.3.6 *Phenomena of Sulphuration*

Some particular aspects concerning sulphuration phenomena should be carefully evaluated when speaking of kinetics and morphology of tinplate corrosion [21].

With concern to thermal sterilisation of canned products such as pulses, meat and fish, sulphuric compounds suffer partial thermal degradation, giving rise to secondary products of decomposition, including hydrogen sulphide (H_2S). In particular, sulphuric compounds are proteins containing sulphur, shortly sulphoproteins.

H_2S, initially localised in the headspace, tends to diffuse all over the inner surface of the can. H_2S reacts with Sn and Fe of the tinplate coating causing sulphur stains and sulphurations. The former ones are made up of tin sulphides and the latter ones of iron sulphide.

The colour of streaks varies from yellow to brown, with light blue and violet iridescences; they can also be not uniform on the inner surface, but more marked if there is more contact between the canned product and the tinplate.

Sulphurations have blackish colour and a spongy and incoherent aspect. They are formed near some pores on tin coating (holes, abrasions) by the reaction of H_2S with steel base.

Should discontinuities of the tin coating have only a standardised porosity, iron sulphurations would obstruct pre-existing holes obstructing the progression of the attack. Should discontinuities be made up of abrasions and fractures, two different cases could happen:

- The product itself prevents iron sulphurations from spreading outside the area of formation, if the product is solid (e.g. meat and fish)
- Sulphuration continues until H_2S is exhausted, if the product is packaged with its brine (e.g. pulses). Brine has a peculiar action of mechanic removal on iron.

From a hygienic or taste viewpoint, sulphuration phenomena on tinplate are not important, but they can represent a problem of 'aesthetic good looking' of the packaging. As a result, the necessity of avoiding similar defects imposes the application of an adequate internal lacquering on surfaces, using specifically formulated paints with a high protective power against sulphuration phenomena.

Table 6.13 Pack test results of a sulphurated product

Fresh Product			Time = 0 month		Time = 1 month		Time = 2 months		Time = 4 months	
Sn (mg/Kg)	Fe (mg/Kg)	Sample	Sn (mg/Kg)	Fe (mg/Kg)	Sn (mg/Kg)	Fe (mg/Kg)	Sn (mg/Kg)	Fe (mg/Kg)	Sn (mg/Kg)	Fe (mg/Kg)
< 5.0	10.88	E2.8 A	10.0	9.98	< 5.0	9.51	< 5.0	9.98	< 5.0	12.77
		E2.8 B	< 5.0	10.64	< 5.0	7.89	< 5.0	10.50	< 5.0	11.24
< 5.0	11.42	E2.8 C	< 5.0	9.85	< 5.0	9.36	< 5.0	9.93	< 5.0	12.95
		E2.8 D	–	–	< 5.0	10.08	< 5.0	10.22	< 5.0	11.24
< 5.0	11.05	E1.4 A	< 5.0	8.73	< 5.0	9.43	8.5	10.35	< 5.0	13.54
		E1.4 B	< 5.0	9.94	< 5.0	9.37	8.4	9.91	< 5.0	14.31
		E1.4 C	< 5.0	9.60	< 5.0	8.85	6.4	12.56	< 5.0	12.73
		E1.4 D	–	–	< 5.0	8.96	5.7	11.54	< 5.0	14.11

Table 6.13 shows results of a pack test on canned beans. After 4 months at 37 °C, the concentration of iron is still very low, similarly to the initial one, confirming that the phenomenon has prevalently an aesthetic—rather than corrosive—effect.

6.3.7 Inhibitors of Corrosion

There are certain substances that accelerate and, in any case, increase polarisation acting, therefore, in a completely different way if compared to depolarisers.

A very important class of these compounds is represented by organic inhibitors. These substances can be chemically absorbed both on the anode and on the cathodic areas at the metal–liquid interphase inside the can, creating insoluble compounds and/or blocking the formation of H_2 coming from the attack of the tinplate.

In particular, some organic anionic compounds have been subject matter of research on the inhibition of corrosion by nitrates; sodium dodecyl sulphate can be mentioned because of the high inhibiting power. In addition, some natural substances, such as agar-agar, gelatin and pectin, form colloidal solutions and act as corrosion inhibitors for superficial absorption. Finally, some spices, in particular garlic and onion, have a passivation effect, which are only kept in the course of time in rather pushed vacuum conditions.

The passivation effect can be linked to sulphur and allyl disulphides. Board et al. [22] have studied the influence of allylthiourea, carbon disulphide and diphenylthiourea on the corrosion rate of tinplate cans filled up with pH 4.00 citrate buffer through electrochemical and packaging tests. The influence of a corrosion inhibitor such as the essential onion oil is evident from the comparison of the values of corrosion rate of D11.2 plain tinplate when immersed in pH = 4.00 model citric solution with decreasing onion essential oil concentrations. The higher the concentration of onion (0.2 %) is, the lower the corrosion rate is (Table 6.14).

Table 6.14 The influence of a corrosion inhibitor such as essential onion oil on corrosion rates of D11.2 plain tinplate. Measured I_{corr} values by means of electrochemical tests

Sample	I_{corr} (µA/cm2)
Citric model solution	4.23
0.20 %—Onion	0.07
0.10 %—Onion	0.51
0.05 %—Onion	1.60
0.01 %—Onion	7.86

6.4 Use of Tin-Free Steel in the Industry of Containers for Canned Foods

Chrome plate (ECCS), also named TFS, has already been a material of standard production in the steel maker industry for several years [9]. Should TFS be used as material for can ends and deep-drawn bodies, the application of a lacquer on both sides would be needed. This necessity is based on some specific features of the material, which can be briefly explained. Firstly, the coating—metallic chromium and chromium oxide—does not offer any electrochemical protection to steel base even if in different proportions depending on the product and the process. On the other side, this coating offers a protection of passive nature, whose efficacy is based on the thickness and uniformity of the same coating. It has to be noted that this coating inevitably has discontinuities because of the low thickness of ECCS.

The necessity of coating is also due to two features of chromium layers: hardness and fragileness. Therefore, the coating is easily damaged on surfaces when in contact with mechanical parts of can-making machines in action. Moreover, it should be needed to observe that TFS or ECCS is characterised by an excellent adherence to coatings and by an excellent workability, particularly with fast machines.

The use of TFS is widespread, and apart from some exceptions, it does not cause any problem. It can be used for weakly acid canned foods without anthocyanins (tomatoes, white fruit) and for non-acid products containing sulphuric compounds (pulses, meat, fish). On the other hand, TFS cannot be used in the presence of canned foods containing more than 1 % of acetic acid or lactic acid.

6.4.1 Aluminium in the Packaging of Canned Foods

Aluminium, used for deep-drawn and redrawn cans in food packaging, is used as the main component of appropriate alloys with other metals in order to obtain optimal mechanical characteristics and the attitude to deep-drawing in several formats.

The spontaneous reaction of Al with atmospheric oxygen leads to the formation of a thin passivated film that gives just a slight protection to corrosion phenomena. Because this film is thin and also not homogeneous, a chemical or electrochemical passivation is produced on industrial lines. However, the oxide layer on the aluminium surface is not a complete protection to the metal because it is removable both at low (<4.0) and at high (>8.0) pH values. In addition, this oxide layer is a porous coating and consequently permeable to many ions. The surface has to be protected by a lacquer in order to improve corrosion resistance.

Laminates of aluminium alloys for deep-drawn cans and lids must always be coated on both sides, internal and external, in order to produce them without forming abrasions. Moreover, these laminates gave to be protected by the contact with canned foods both in the liquid state, or containing liquids that are more or less aggressive, and in the dry state. The main cause is the abrasive action that they can carry out during the transport.

The choice of lacquer and enamels is determined by two main factors:

- Ratio of deep-drawing (deep-drawn cans into two pieces)
- Aggressiveness of the food product. The level of porosity of the coating in the internal part of the can caused by lengthening, owing to aluminium stretching during the deep-drawing, must be adequate to the aggressiveness of the food product.

After this premise, three groups of products must be taken into account. Data and information reported are essential examples.

– Non-aggressive products (e.g. Pâté, meat jelly, pudding); pH = 5–6, salt < 2 % and/or greasy substances
– Medium aggressive products (fish in tomato sauce, meat sauce, etc.); pH = 4.7– 5.0, salt about 2 %, without greasy substances and oil
– Aggressive products (also containing acetum); pH = 3–5, salt = 2 %, acetic acid ≤ 0.3 %, oils: 3–5 %, citric acid.

All products that are canned in tinplate containers can also be packaged in aluminium cans.

Cans made up of aluminium alloy are not subject to phenomena of sulphuration with products containing sulfurated proteins because aluminium sulphide that could be formed hydrolyses creating H_2S and aluminium oxide. They do not give place to undermining corrosions.

Corrosive attacks can take place with acid products in correspondence of pores of the lacquer in more stressed areas (attacks are delayed by the layer of oxide of passivation); during the attack, there is the development of H_2 that causes a decrease in the degree of vacuum and possible swelling of the can. The most aggressive component of food products for aluminium is sodium chloride. An example of the influence of the type of product on the corrosion rate of aluminium for drink cans is shown in Table 6.15.

Another example of corrosion of lacquered aluminium is referred to easy-peel lids in contact with very acid vegetal product, for example citric acid, acetic acid or lemon juice (pH = 3.3–3.6). Some batches of aluminium packs filled with different types of vegetables and fruits puree swollen after few months of shelf life at room temperature. After few months of storage, Al concentration in the product

Table 6.15 The influence of the type of product (tea, cola, wine) on the corrosion rate of aluminium beverage cans

	Tea sample	Cola sample	Wine sample
Al dissolved under steady state conditions after 17 days (mg)	2.62	0.17	2.08
(mg/dm^2)	13.10	0.85	10.40
Electrochemical corrosion rate (μA/cm^2)	1.15	0.20	0.65
Visual examination	Development of a passivation film: slight uniform corrosion	Development of a passivation film	Localised corrosion: some pits

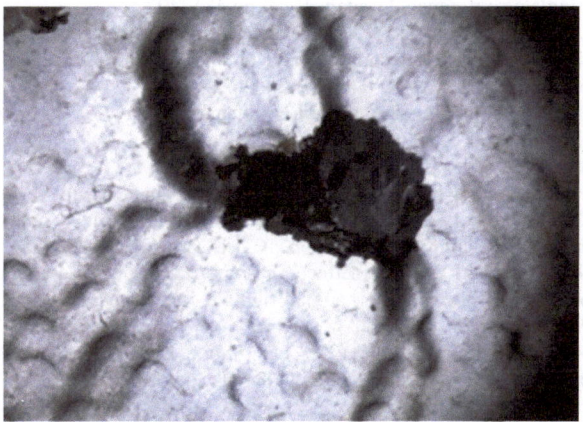

Fig. 6.9 Examples of corrosion and perforations in lacquered aluminium: easy-peel lids in contact with acid vegetable products (pH = 3.3–3.6). After a few months of storage, the concentration of aluminium in food products may arrive to 88.7 mg/kg. At the same time, surfaces of lids can show different perforations

has been detected up to 88.7 mg/kg and the surface of lids has shown perforation in different areas as shown in Fig. 6.9.

References

1. Hinds JTG (2012) The electrochemistry of corrosion. National Physical Laboratory, Teddington. Available http://www.npl.co.uk/upload/pdf/the_electrochemistry_of_corrosion_with_figures.pdf. Accessed 07 Nov 2014
2. Shreir LL (1963) Corrosion, vol 2. George Newnes Ltd, London
3. Fontana MG, Greene ND (1967) Corrosion Engineering. McGraw-Hill Book Company, New York
4. Marsal P (1989) The Can and its uses, Part2. The Canmaker, Crawley, p 59

5. Montanari A, Milanese A (2001) Materiali metallici e contenitori per l'industria alimentare. Collana Monografie SSICA, Stazione Sperimentale per l'Industria delle Conserve
   Alimentari, Parma
6. Massini R (1973) La corrosione della banda stagnata da parte di conserve alimentari—I: ele-
   menti generali di teoria elettrochimica dei processi di corrosione. Ind Conserv 48(4):237–245
7. Morgan E (1985) Tinplate and Modern Canmaking Technology. Pergamon Press, Oxford
8. UNI EN 10202:2004. Cold reduced tinmill products—electrolytic tinplate and electrolytic
   chromium/chromium oxide coated steel. Ente Nazionale Italiano di Unificazione (UNI),
   Milan
9. Ferrari F, Pacelli L, Montanari A, Cassarà A, Riccio M (1991) Main properties and performances of tin free steel CT. Paper presented at the second high current density ECCS
   Conference, Genoa, May 1991
10. Dong SL, Kit LY, Piergiovanni L (2008) Food packaging science and technology. CRC Press,
    Boca Raton
11. Massini R, Montanari A, Milanese G, De Anna PL (1984) Improvements in electrochemi-
    cal techniques for evaluating the corrosion behaviour of tinplate. In: Proceedings of the third
    international tinplate conference, London, 1984, pp 481–492
12. Montanari A, Pezzani A, Cassara A, Quaranta A, Lupi R (1996) Quality of organic coat-
    ings for food cans: evaluation techniques and prospects of improvement. Progr Org Coat
    29(1):159–165. doi:10.1016/S0300-9440(96)00625-X
13. Turner TA (1998) Canmaking. The technology of metal protection and decoration. Blackie
    Academic & Professional, London
14. Larousse J, Brown BE (eds) (1997) Food canning technology. Wiley-VCH, New York
15. Kamm GG, Hotchner SJ, Kopetz A (1988) Corrosion anomalies with light-coloured fruit in
    tinplate cans produced from aluminium- killed continuous cast-steel. In: Proceedings of the
    4th international tinplate conference, London, 10–14 Oct 1988, pp 356–381
16. Nishida H, Ogha T, Oyagi Y (1992) Effects of composition of can contents on rapid detin-
    ning. Paper presented at the 5th international tinplate conference, London, 12–16 Oct 1992,
    No 21, pp 1–9
17. Palmieri A, Montanari A, Fasanaro G. (2004) De-tinning corrosion of cans filled with tomato
    products. Corros Eng Sci Technol 39(3):198–208. doi:http://dx.doi.org/10.1179/1478422
    04X2808
18. Zurlini C, Montanari A, Squitieri G, Gelati S (2010) Shelf-life study of lacquered canned
    tomato: influence of different variables of the process/product. Paper presented at the asian
    steel packaging conference, Kuala Lumpur, 23–24 Sept 2010
19. Montanari A, Marmiroli G, Pezzani A, Cassarà A, Lupi R (1995) Easy open ends for food
    cans: definition, organic coatings and problems involved. In: Proceedings of the international
    congress of paints, pigments, varnishes, printing inks and adhesives, EUROCOAT 95, Lyon,
    19–21 Sept 1995
20. Barella S, Cincera S, Boniardi M, Bellogini M, Gelati S, Montanari A (2011) Failure analysis
    of tuna cans. J Fail Anal Prev 11(4):446–451. doi:10.1007/s11668-011-9464-x
21. Montanari A, Pezzani A, Cassarà A, Lupi R, Rocchi P (1994) Sulphur-stain and corrosion
    resistance of metal food cans coated with zinc-rich lacquers. In: Proceedings of the interna-
    tional conference UK CORROSION and EUROCORR 94, Vol 1, Bournemouth, 31 Oct–03
    Nov 1994
22. Board PW, Holland RV, Elbourne RGP (1967) The effect of sulphur-containing fungi-
    cides on the corrosion of plain cans of fruit. J Sci Food Agric 18(6):232–236. doi:10.1002/j
    sfa.2740180603
23. Montanari A, Milanese G, Cassara A, Tomasicchio M, Barbieri G, De Giorgi A, Pezzani A
    (1992) Corrosion problems of tinplate for artichoke packs. In: Proceedings of the fifth inter-
    national tinplate conference, London, 1984, pp 196–221